SHORT-CUT
MATH

SHORT-CUT MATH

Gerard W. Kelly

DOVER PUBLICATIONS, INC.
New York

This Dover edition, first published in 1984, is an unabridged
and slightly revised republication of the edition first published
by Sterling Publishing Co., Inc., N.Y., in 1969. A few correc-
tions have been made in the text and the Introduction has been
slightly expanded.

Library of Congress Cataloging in Publication Data

Kelly, Gerard W.
    Short-cut math.
    "An unabridged and slightly revised republication of the
edition first published by Sterling Publishing Co., Inc., 1969"
—T.p. verso.
    Includes index.
    1. Ready-reckoners. I. Title.
QA111.K44 1984   513'.92   83-20634
ISBN-13: 978-0-486-24611-6
ISBN-10: 0-486-24611-6

Manufactured in the United States by Courier Corporation
24611634
www.doverpublications.com

CONTENTS

MATH AS YOU LIKE IT

Do you yearn for multiplication
You can do without frustration?
Does it please you not to carry in your head?
Do you wish some court decision
Would abolish long division?
Do you look upon remainders with some dread?

Does the borrow in subtraction
Often drive you to distraction?
Do you dream of doing fractions in a breeze?
With each answer must you fidget
To be sure of ev'ry digit?
Wouldn't you rather do it all with speed and ease?

INTRODUCTION

This book contains about 150 short-cuts that provide faster, easier ways to add, subtract, multiply and divide. Anyone who uses numbers and does calculations can benefit from these short methods and even enjoy using them. You can apply them at school, home, shop, or office to save time and trouble. Short-cuts can help to improve your math and provide the urge to learn more or to review the math you have forgotten. Or you can use short-cuts simply because they are fascinating and make it fun to work with numbers.

The short-cuts in this book require no special ability in mathematics. If you can do ordinary arithmetic, you will have no trouble in using these methods. There is nothing to "unlearn," no foreign devices, strange systems, or difficult operations that would require long study and practice. Furthermore, the book uses only a minimum of technical terms, so you don't have to relearn what an augend, a minuend, or a multiplicand is, in case you have forgotten your school definitions.

To get the most out of this book, go through it briefly to see what it covers and how it is organized. After that, study whatever methods are of special interest to *you*. Naturally, you will not be able to use, or even to remember, every short-cut (to remember a short-cut is usually the hardest part of it). The best approach is to concentrate on the methods you can use in your daily work or the ones you would like to learn just for the fun of knowing them.

Examples are given under each short-cut to demonstrate how it works. However, there are no long drills and exercises. You will still need to practice any short-cut you want to put to work but you will probably find it more satisfying to make up your own problems, keeping in mind that the best way to develop speed, ease, and accuracy with any method is to use it regularly.

Among the features used throughout the book are "Proving the answer" and "Why it works." Proof of the answer is given to demonstrate that the short-cut really arrives at the same answer as ordinary arithmetic, a result often so surprising that you may find it hard to believe. Proving the answer also helps to show how much easier the short-cut is and to encourage the good habit of checking results.

"Why it works" is included under many short-cuts where a short, simple explanation can be devised. This aids in understanding the method and in remembering it, too. Even if you forget the details of a short-cut, you can often reconstruct the method by recalling the "why" of it and working out an example with small numbers. This will give you a simple pattern to follow and will make the short-cut easier to remember and use.

Under each short-cut in this book, either a rapid "Solution" or a longer "Solution by steps" is worked out. The steps are used where needed for additional clarity and are followed by a "Short way" indicating how the method would actually be used.

Also, after a group of related methods, a summary or table is given so you can see the methods at a glance, and as a further help in remembering them.

As you work with short-cuts, you will probably see other ways of calculating faster and easier. Discovering methods of your own can add new excitement to your journey through this corner of the magic world of numbers.

The short-cuts, of course, are not intended to supplant the pocket calculator. Rather, there's a place for both. For long, complicated computations the electronic marvel has no rival, but for simpler problems a short-cut or rapid mental calculation can often give the answer faster than you can reach for the machine. And don't forget those old reliables, pencil and paper. They're inexpensive, easily carried, never need a battery, and give you a record of what you did so the work can be checked. Each of these has advantages of its own, so choose whichever method or combination of methods best suits your need and the problem to be done.

If you use a calculator, short-cuts can provide an easy way to estimate the answer in advance. This should be done to guard against large errors since pushing some wrong keys can send the answer into Never-Never Land; and since the machine leaves no record of entries, there's no way of telling what keys you *did* push. An estimate helps to warn of big errors that might otherwise go unnoticed.

And, of course, a calculator can't run itself. It needs *you* to tell it what to do and this requires a knowledge of arithmetic. It is hoped that this book will assist toward a better understanding of the basics involved and a deeper appreciation of the wonders of working with numbers.

I. FUNDAMENTALS OF
SHORT-CUT METHODS

1-1. Basic Ways of Simplifying Calculations

Short-cut methods are based upon the principle of changing "difficult" numbers and processes into easier ones. For example, to find the sum of 29 and 36, you can add 1 to 29 and subtract 1 from 36 so that the problem becomes $30 + 35$; you can see at once that the answer is 65. Adding and subtracting 1 simplified the problem, but did not change the answer. This principle of "equivalency" as used in short-cuts enables you to reach the right answer faster and easier.

You can apply the same principle to multiplying. Take 28×15, for instance. If you halve 28 ($= 14$) and double 15 ($= 30$), you change the problem to 14×30 which, as you can see, is an easy 420—the same answer you will get by multiplying 28 by 15.

It's not always the *size* of a number that makes it difficult to handle but the KIND of number it is. You know, of course, that it's much easier to multiply by 10 than by 9 even though 10 is the larger number. However, the "difficult" number 9 can be changed into two "easy" numbers, 10 minus 1, which equal 9. Therefore, to multiply a number by 9, you can multiply by 10 (simply add a zero to the number) and subtract the number, thus: $37 \times 9 = 370 - 37 = 333$. The $370 - 37$ is, of course, $37 \times (10 - 1)$ instead of 37×9. Either way, the answer is the same.

Incidentally, the easiest numbers to handle are 0, 1, 10, and 2, although 100, 1,000, etc., are also easy to get along with; so are other "zero numbers," such as 20, 30, etc. If you can change a difficult number to one of these, you are on your way to an easier solution.

In many cases, you will find it faster and easier to perform two or three simple operations rather than a single more difficult one, like taking the stairs to the next floor rather than trying to make it in one big jump. To multiply 64 by 25, for instance, you can divide 64 by 4 ($= 16$) and multiply by 100 (just add two zeros), and there's your answer: 1,600. This works because dividing by 4 and multiplying by 100 is the same as multiplying by $\frac{100}{4}$ which equals 25. Again, the method has been simplified but the answer is not changed because 25 and $\frac{100}{4}$ are equivalent.

Another basic way of simplifying a calculation is to break up the numbers into easier parts. For example, to divide by 16, you can divide first by 2 and then by 8, or divide twice by 4, since 2×8 and 4×4 are each equivalent to 16. Try dividing 16 into a number like 224. You can divide the number in half ($= 112$) and divide this by 8 ($= 14$), or you can divide 224 by 4 ($= 56$) and divide again by 4 ($= 14$). Dividing 224 by 16 will, of course, give the same answer, 14, because all three ways of dividing are equivalent.

A number may be broken up into other parts, too, such as into units, tens, hundreds, etc. These smaller parts can make the number much easier to handle. The number 613, for instance, can be broken up into $600 + 10 + 3$. If you want to multiply 613 by another number, such as by 12, you can multiply each part and then add the partial results to get the answer, thus:

$$
\begin{array}{rcl}
600 \times 12 & = & 7{,}200 \\
10 \times 12 & = & 120 \\
\underline{3 \times 12} & = & \underline{36} \\
613 \times 12 & = & 7{,}356
\end{array}
$$

Besides such methods as the foregoing, short-cuts are also based upon eliminating or easing processes that most people find difficult, such as "carrying" in addition and multiplication, "borrowing" in subtraction, the intricacies of "long" division, and the need to remember results when doing problems mentally.

1-2. Results Are What You Want

In doing a problem, it's better to think only of the *results* of each step and to omit any unnecessary details of the step itself. For example, when you see $36 \div 9$, or $9\overline{)36}$, think immediately of 4. Don't go through the details of the process such as by saying "36 divided by 9 goes 4 times." Or when adding, like $5 + 11 + 8$, you just slow yourself up if you say "5 plus 11 makes 16, plus 8 makes 24." Instead, just think: "16, 24."

Omit the words of a process as much as possible and concentrate on the results. You can work problems faster without the words.

Numbers and calculating are like the alphabet and reading.

When you see a word, you don't stop to spell out the letters that make it up; you recognize the word at once in its entirety. Nor do you stop to consider each word separately; you read words together and think of what they mean as a whole. Doing the same with numbers and calculations will enable you to solve problems faster and easier.

1-3. Combining Mental and Written Math for Best Results

Short-cuts are not just a way of solving problems in your mind and in a flash, although they *can* help you do this. Taking the time to jot things down can often speed them up. Consider the following:

Calculating involves the mind in two processes: (1) thinking out each step in the problem, and (2) remembering the results of each step for use in succeeding steps. Naturally, the longer the problem and the more complex, the more you have to remember if you do it entirely by mental math, and the more chances there are of making a mistake.

On the other hand, if you work out the whole problem on paper, you don't have to keep remembering the results of steps and you can concentrate on the solution. Writing it down may take longer but it reduces the likelihood of error, and this can save time in the long run.

You can often achieve the maximum of speed with accuracy by using both the mental and the written methods. You do this by writing down the *results* of the various steps but not the details of the steps. For example, to multiply 49,837 by 11, there is no need to write out the whole problem or to struggle with it as a mental exercise; just write the partial products, then add to get the answer, thus:

$$
\begin{array}{r}
49837 \\
\underline{49837} \\
548207
\end{array}
$$

Or, if you want to divide 2,184 by 24, you might divide by 3 and write down 728 so you don't have to remember it; then divide by 8 to get the answer, 91.

Some short-cuts in this book will enable you to solve problems mentally, and even on sight; others will work better if you write down partial results by steps; still other short-cuts will serve you best if you do the problem entirely on paper. It's up to you to determine how much to carry in your head and how much to write down in order to achieve both speed and accuracy. While speed is desirable, a short-cut to the wrong answer is obviously useless. It pays to take a longer route, if necessary, to reach the right answer the first time.

1-4. Approximating and Rounding Off

Before working out a problem, determine how accurate your answer must be. A good estimate or approximation may be all you need and can save you a lot of needless figuring. Approximating depends upon determining the "significant figures" of the numbers involved—the figures that are important to your estimate.

For example, if an item is priced at 2 for $1.97, you can't go far wrong in figuring that 3 will cost about $3.00. Or if you want to paint some walls measured at 2,386 square feet, you might use 2,500 in estimating how much paint you will need. If a gallon of paint covers 500 square feet, you have your answer faster than you can reach for a pencil.

To arrive at the significant figures, "round off" a number to whatever extent may be required to suit your purpose and drop all figures (digits) to the right of this. If the first digit dropped on the right is 5 or more, add 1 to the last digit remaining; otherwise let the last digit stand as it is. The following examples will show how this is done.

Example (1): Round off 6,437:
To the nearest thousand: 6,000
To the nearest hundred: 6,400
To the nearest ten: 6,440

Example (2): Round off 83.652:
To the nearest unit: 84
To the nearest tenth: 83.7
To the nearest hundredth: 83.65

Ways for rounding off numbers to approximate or estimate the answer in addition, multiplication, and division are given in the chapters dealing with these operations.

1-5. Achieving Accuracy with Speed

Some short-cuts promote both speed and accuracy because they enable you to see the correct answer at once. However, where several steps are involved, and especially with large numbers, there is much more chance of error. As previously suggested, you will often save time in reaching the right answer if you write down at least some of the results as you go.

You can further ensure accuracy by estimating your answer in advance and then checking the answer after you get it, as discussed next.

Estimating the result in advance:

Estimating the answer before doing any problem helps to avoid big mistakes, such as omitting a decimal point or putting it in the wrong place, or writing 14,444 instead of 1,444, etc. An estimate can catch errors that make the answer much larger or smaller than it should be. You can estimate your answer by rounding off numbers or by using a short-cut, as covered in later chapters.

Checking your answer:

You can check your answer by doing the problem again, either in reverse order or by some different method. Doing the problem in a different way helps to avoid the repetition of a mistake that may be habitual.

Various ways for checking your answer in addition, subtraction, multiplication, and division are given in the chapters dealing with these operations. These checks include "casting out 9's" which can be used to test all operations. The process of casting out 9's is explained in 1-6 (the next section).

However, no method of checking gives absolute proof that your answer is correct. There is always the chance that you might make a mistake in the check and confirm a wrong answer. However, if your answer does check out, you can usually assume it is correct.

Another essential for achieving accuracy with speed is to keep numbers and decimal points properly placed in every problem. The careful alignment of digits by units, tens, hundreds, etc., will help to attain fast, accurate results.

And finally, accuracy and speed also require that you be reasonably well skilled in basic arithmetic and can quickly recognize that $9 + 8 = 17$, $16 - 5 = 11$, $8 \times 6 = 48$, $56 \div 7 = 8$, etc. If you must stop to work out problems like these, you will need to review and practice your "tables."

1-6. Testing Answers by Casting Out 9's

This test is based upon a remarkable property of 9. When any number is divided by 9, the remainder will equal the sum of the number's digits, or the sum of the number's digits after "casting out" 9's. This may sound strange but the idea itself is easy to understand, as the following examples will show.

Example (1): $16 \div 9 = 1$, with 7 remainder. The digit sum of

16 (1 + 6) is also 7, the same as the remainder after dividing 16 by 9.

Example (2): 24 ÷ 9 = 2, with 6 remainder. The digit sum of 24 (2 + 4) is also 6.

Example (3): 38 ÷ 9 = 4, with 2 remainder. Of course, the digit sum of 38 is not 2 but 11. However, after casting out 9, such as by subtracting 9 from 11, the remainder does become 2, the same as the remainder after dividing 38 by 9.

Remainders obtained by casting out 9's can be used to provide "check numbers" for testing your answers in addition, subtraction, multiplication, and division.

You can use any of four methods to cast out 9's and obtain the remainders, as shown in the following examples. The same numbers are used in each example to demonstrate that the remainders obtained are the same for all four methods.

A. Casting out 9's by dividing:
Divide the number by 9 to obtain the remainder.

Number	Remainder
46	1
85	4
198	0
6,368	5

B. Casting out 9's by adding:
Add the number's digits, and add the digits in the sum if there is more than one digit in the sum, to obtain a one-digit remainder. If the remainder is 9 or a number evenly divisible by 9 (18, 27, 36, etc.), count the remainder as zero.

Number		Remainder
46:	4 + 6 = 10; 1 + 0 =	1
85:	8 + 5 = 13; 1 + 3 =	4
198:	1 + 9 + 8 = 18; 1 + 8 = 9;	0
6,368:	6 + 3 + 6 + 8 = 23; 2 + 3 = 5	

C. Casting out 9's by subtracting:
Add the number's digits and subtract 9 each time the sum equals or exceeds 9. Do this until the final remainder becomes less than 9. Or, you can first get the digit sum and then keep subtracting 9 until the remainder becomes less than 9.

	Number		Remainder
	46:	$4 + 6 = 10; - 9 =$	1
	85:	$8 + 5 = 13; - 9 =$	4
	198:	$1 + 9 = 10; - 9 = 1; + 8 = 9; - 9 =$	0
also:	198:	$1 + 9 + 8 = 18; - 9 = 9; - 9 =$	0
	6,368:	$6 + 3 = 9; - 9 = 0; + 6 + 8 = 14; - 9 =$	5
also:	6,368:	$6 + 3 + 6 + 8 = 23; - 9 = 14; - 9 =$	5

D. Casting out 9's by omitting them:

Add the number's digits but do not include 9's or combinations equal to 9 (6 and 3; 5 and 4; 1, 3, and 5, etc.). Omitting 9's and 9-combinations saves adding them up and subtracting 9's later. If the sum comes to 9 or more, cast out 9's to obtain the remainder.

	Number	Remainder
46:	no 9 to omit; $4 + 6 = 10; - 9 = 1$	
85:	no 9 to omit; $8 + 5 = 13; - 9 = 4$	
198:	omit $1 + 8$ and omit 9; $=$	0
6,368:	omit $6 + 3$; $6 + 8 = 14; - 9 =$	5

Did you notice that the remainders were the same for all four methods?

Omitting 9's is the easiest way to cast them out, especially with large numbers, since it reduces the adding and subtracting, and requires no division. In a number like 2,756,942, for instance, the 2 and 7, 5 and 4, and the 9 can be omitted, leaving only 6 and 2 to be added, giving a remainder of 8. Dividing 2,756,942 by 9 will, of course, also give a remainder of 8, as shown:

$$9)\ \overline{2,756,942} \\ 306,326\tfrac{8}{9}$$

How you can apply casting out 9's to quickly check the results of adding, multiplying, and dividing is shown at the end of the chapters dealing with these operations. The method does not give positive proof that your answer is right (no method does) but if your answer does not meet the test of casting out 9's, you can be sure that there is an error somewhere, either in doing the calculation, or in making the check, or in both.

II. ADDITION

2-1. Basic Principles

Numbers may be added in any order. For example, $7 + 4 + 3 = 14$; $3 + 4 + 7 = 14$; $4 + 3 + 7 = 14$, and so on.

You can take advantage of this principle by adding numbers in whatever order is easiest for you. In getting the sum of $99 + 83 + 1$, for instance, you can first add 99 and 1 to make 100, and then add 83 to get 183. Obviously, this is easier than adding 99 and 83, and then 1.

You can also simplify addition by changing difficult numbers into easier ones. For example, to add 39 and 47:

(*a*) Simplify 39 into a number ending in zero: $39 + 1 = 40$

(*b*) Add: $40 + 47 = 87$

(*c*) Subtract the 1 you added: $87 - 1 = 86$, answer

Or you can do it this way:

$(39 + 1) + (47 - 1) = 40 + 46 = 86$

By adding an amount to one number and subtracting an equal amount from a second number, the sum of the two numbers remains the same.

As mentioned in Chapter I, you can do problems faster if you cut down on words as much as possible. In adding $9 + 4 + 7$, for example, you would not say "9 and 4 make 13, and 7 makes 20." Instead, you "think" the addition without the words: 13, 20. You would do the same when adding a column.

```
 5
 9
 4    Omit the words and think only of the result of each step:
 3    15, 18, 22, 31, 36.
 7
 8
──
36
```

You should be able to "see" the sum of two digits at once,

without stopping to count, and this means that you must be on good terms with your addition table, at least up to 9 + 9.

2-2. Adding by 10-Groups

Adding by 10 is easy, so look for combinations that add to 10.

```
 ⎛4
 ⎜5⎞   Notice the three pairs of numbers that make 10-groups:
 ⎝6⎠   2, 8; 5, 5; and 6, 4. These can be added quickly in three
  5⁄   steps: 10, 20, 30.
 ⎛8
 ⎝2
 ──
 30
```

Pairs of numbers that add to 10:
1, 9; 2, 8; 3, 7; 4, 6; 5, 5

Trios of numbers that add to 10:
1, 1, 8; 1, 2, 7; 1, 3, 6; 1, 4, 5
2, 2, 6; 2, 3, 5; 2, 4, 4
3, 3, 4

Since numbers may be added in any order, it makes no difference whether a pair appears as 7, 3, or 3, 7; or a trio as 1, 3, 6, or 3, 6, 1, or 6, 1, 3, and so on. They all make an easy 10 to add.

Of course, you can extend the same idea to adding by 20-groups: 2, 9, 9; 3, 8, 9; 4, 7, 9; 4, 8, 8, etc. By recognizing such combinations and converting them to a single number ending in zero, addition is made easier.

2-3. Reaching 10-Levels

In adding, look for opportunities to reach a 10-level (20, 30, 40, etc.). For instance, if you have added up to 36, look for a 4, or a 3 and a 1, or a pair of 2's, to make an easy 40; or if your addition is at 62, look for an 8 or some 8-combination that will take you to 70.

Because a 10-level number ends in zero, you will find it easier to remember the sum as you go and easier to add the next number.

2-4. Adding by Other Combinations of Numbers

Besides combinations that make 10 or 20, you can add other numbers by pairs or trios as though they were single numbers, such as adding 4 and 3 as 7, 5 and 1 as 6, 1, 3, and 5 as 9, etc. If you learn to "see" these groups as a single number, you will be able to add faster and easier.

```
 21
 14   Starting from the bottom, addition by groups could go
 33   like this: 9, 19, 24; write 4, carry 2; 10, 23, 29.
 47
 96
 83
294
```

2-5. Adding by Multiplying

Multiplication is simply a rapid way of adding the same number over and over. For example, 7×4 is the same as adding 7 four times: $7 + 7 + 7 + 7$. The result comes to 28 either way.

If a number is repeated, especially if repeated often, you can use multiplication to speed up your addition. After multiplying, add the other numbers.

```
 8
 7
 5   8 × 3 = 24; 30, 42
 8
 6
 8
42
```

2-6. Adding Two or More Columns

In writing columns of numbers for addition, line up the columns carefully so that units are under units, tens under tens, and so on. If decimal points are used, keep these in a straight column, too. An orderly arrangement helps to promote both speed and accuracy.

As you get the sum of each column, write the "carry" as a small number above the next column on the left.

```
 32
 346
 521
 789   The sum of the units column is 27; write 7, carry 2; the
 640   sum of the tens column is 31; write 1, carry 3.
 375
 246
2,917
```

You may prefer to write the carry above and to the side of the numbers in the answer, such as 2,9$\overset{32}{1}$7. Either way, writing the carry relieves your mind of remembering what it is and makes it easier to locate an error if a check shows that the answer is wrong. Since the written carry lets you see the total for each column, you can check each column separately. Furthermore, if you are interrupted while adding, you will not have to re-add any column where the carry is shown.

2-7. Adding Long Columns

In a long column, you can add the numbers by groups to get sub-totals and then add the sub-totals to get the sum total. This gives your mind a chance to rest, makes it easier to check your answer, and enables you to locate a mistake more readily in case there is one. As a further help, you can write the carry for each group in the sub-totals.

```
4,382
2,654
7,655
9,872    231
6,341   30,904   Sub-total
9,534
7,228
5,287
6,634    112
9,110   37,793   Sub-total
        68,697   Sum total
```

2-8. No-Carry Addition

You can avoid carrying in addition by writing the total for each column, then adding the column-totals to obtain the sum total.

```
6,274
4,633
7,452
3,516
8,775
1,942
   22    Notice that the column-totals 22, 27, 33, and 29 are
   27    lined up in their proper places. They are actually 22,
  3,3    270, 3,300, and 29,000.
 29
32,592
```

```
 6,274
 4,633
 7,452   You can also add the columns from left to right.
 3,516   The zeros in the column-totals are not necessary but
 8,775   are shown to indicate the real values of these totals.
 1,942
29,000
 3,300
   270
    22
32,592
```

Left-to-right addition can be useful in making estimates since you can stop adding after any column that suits the accuracy you need. In the example just given, the sum of the first column on the left shows that the total is at least 29,000; adding the next column brings the total to at least 32,300. If this is close enough for your purpose, you can skip the addition of the next two columns.

2-9. Rounding Off Numbers for Addition

An even better way to estimate a sum is to round off the numbers. For example, suppose you want to estimate the sum of the following numbers to the nearest thousand.

```
46,274   To round off each number to the nearest thousand,
34,633   add 1 to the thousand digit if the hundred digit is 5
57,452   or more; otherwise, let the thousand digit stand as
13,516   it is. The rounded numbers would then be:
88,775              46
41,942              35
282,592             57
                    14
                    89
                    42
                   283 (thousand)
```

The sum of the rounded numbers is only 408 greater than the sum of the original numbers.

2-10. Rounding Off Dollars and Cents for Addition

In adding sums of money, you can approximate the total by rounding off to the nearest dollar and dropping the cents. Cents of 50 or more are counted as a dollar and added to the dollar unit; 49 cents or less are dropped.

Exact Amounts	Rounded Amounts
I II2 I	
$5,621.42	$5,621
3,842.59	3,843
8,034.91	8,035
7,156.21	7,156
$24,655.13	$24,655

This method saves writing and adding two columns of figures. As you can see, the adding and subtracting of cents tends to keep the total about the same, and as the dollar amounts become larger, the importance or significance of the cents becomes less anyway.

2-11. Adding a Regular Series of Numbers

Example (1): What is the sum of all the numbers from 1 to 15?
Solution by steps:
(a) Add the first and last numbers of the series: $1 + 15 = 16$
(b) Divide by 2 to get the average number: $16 \div 2 = 8$
(c) Multiply the average by the number of numbers in the series: $8 \times 15 = 120$, answer

Short way: $\dfrac{1 + 15}{2} = 8$; $\times 15 = 120$

If you add all the numbers from 1 to 15, you will see for yourself that their sum is 120.

Example (2): What is the sum of all the numbers from 20 to 29?

Solution: $\dfrac{20 + 29}{2} = 24\frac{1}{2}$; $\times 10 = 245$, answer

Proving the answer:

20	
21	
22	
23	
24	Notice that there are ten numbers
25	(not nine) from 20 to 29, inclusive.
26	
27	
28	
29	
245	

Example (3): Find the sum of all the numbers from 1 to 99.

Solution: $\dfrac{1 + 99}{2} = 50$; × 99 = 4,950, answer

2-12. Add-and-Subtract Method of Simplifying Numbers

You may be able to simplify an addition by an easy addition and subtraction before you add the numbers.

Example: 97 + 237

Solution by steps:
(a) Add 3 to 97 to make it an easy 100
(b) 100 + 237 = 337
(c) Subtract the 3 you added: 337 – 3 = 334, answer

Alternate solution:
(a) 97 + 3 = 100
(b) 237 – 3 = 234
(c) 100 + 234 = 334

Short way: 97 + 3 = 100; + (237 – 3) = 334
Proving the answer:

$$\begin{array}{r} 97 \\ + \underline{237} \\ 334 \end{array}$$

Why it works: In any addition, adding and subtracting the same amount does not change the sum. Try this with any numbers you please; for example, 6 + 3 = 9. Now add 2 and subtract 2: (6 + 2) + (3 – 2) = 8 + 1 = 9. Or add and subtract 3: (6 + 3) + (3 – 3) = 9 + 0 = 9.

You can apply this same idea to add difficult numbers like 9, 99, 999, etc. Simply add 1 to change the number to an easy 10, 100, 1,000, etc., and subtract 1 from the sum or from the other number.

Example: 13,862 + 999

Solution:
999 + 1 = 1,000; + 13,862 = 14,862; – 1 = 14,861, answer

2-13. Splitting Numbers into Easy Parts

If numbers contain two or more digits, splitting the numbers into smaller parts can make them easier to add.

Example (1): 76 + 57

Solution: Think of 76 as (70 + 6) and 57 as (50 + 7), then add by parts: 70 + 50 = 120; + 6 + 7 = 133, answer

Example (2): 6,614 + 4,238
Solution: 6,614 = 6,600 + 14
4,238 = 4,200 + 38
10,800 + 52 = 10,852, answer

Or you might find it easier to do it this way:
6,000 + 4,000 = 10,000; + 600 + 200 = 10,800; + 10 + 30
= 10,840; + 4 + 8 = 10,852

2-14. Thinking of Numbers as "Dollars" and "Cents"

This is simply another way of splitting numbers into smaller parts for easier addition.
Example: 962 + 837
Solution: Think of the numbers as $9.62 and $8.37, then add the dollars and cents separately:
$9 + $8 = $17; 62¢ + 37¢ = 99¢;
total: $17.99, which, of course, is 1,799

Checking Your Answer in Addition

No method of checking addition can tell for sure that your answer is right. You might make the same error in the check as you made in the original addition. There are such things as habitual mistakes that are repeated when adding certain numbers, like mistakes in regularly misspelling certain words.

Keep in mind that the significance or size of an error increases as the numbers go from right to left—from units, to tens, to hundreds, and so on. In adding money, for instance, an error of 5 in the cents column would only put you in or out a nickel, but the same error four columns to the left would be a 500-dollar mistake!

2-15. Checking Your Answer by Adding Again in Reverse

You can check a sum by adding again in a different order. If you add from the bottom up the first time, add from the top down as a check. If you get the same answer both ways, you may assume that you are right.

2-16. Checking Addition by the No-Carry Method

If you add in the regular way the first time, you can check your answer by getting sums by columns and then adding these to obtain the total. You can also reverse whatever order you used the first time: top to bottom or bottom to top; right to left or left to right.

	Sums by Columns Right to Left	Sums by Columns Left to Right
3 22		
6,274		
4,633		
7,452	22	29
3,516	27	33
8,775	33	27
1,942	29	22
32,592	32592	32592

2-17. Checking Addition by Casting Out 9's

Various ways to cast out 9's are shown in 1-6, Chapter I. How to apply this method to checking addition is discussed here. The longer the addition and the larger the numbers, the more time you will save by checking your answer by casting out 9's.

The remainders obtained after casting out 9's are used as check numbers for testing the answer in addition, as follows:

Example (1):

	251	Remainder:	8
	672	,, :	6
	415	,, :	1
	982	,, :	1
	2,320	Total: 16:	7 Remainder

Remainder of answer (2,320): 7

Since the two remainders (7) are equal, you may assume that the answer is correct.

Example (2):

	729	Remainder:	0
	436	,, :	4
	184	,, :	.4
	655	,, :	7
	2,004	Total: 15:	6 Remainder

Remainder of answer (2,004): 6

Remainders are equal; the answer checks.

Casting out 9's does not give positive proof that the answer is correct but if the answer does *not* check, there is definitely an error in the addition, in the check, or in both.

III. SUBTRACTION

3-1. Introduction

Subtraction is the reverse of addition; unlike addition, it cannot be done in any order. $5 + 3 = 8$ and $3 + 5 = 8$ but obviously $5 - 3$ and $3 - 5$ are not the same.

"Borrowing" is usually regarded as the main difficulty in subtraction. There are some ways to eliminate or reduce this but there are not as many short-cuts in subtraction as in the other operations.

Subtraction shows the *difference* or *remainder* after taking one number from another, and may be written in either of the following ways:

$9 - 6 = 3$ Remainder

$$\begin{array}{r} 9 \\ -\underline{6} \\ 3 \text{ Remainder} \end{array}$$

3-2. Add-or-Subtract Method of Simplifying Numbers

You can simplify subtraction by changing the number to be subtracted into a number ending in zero. This can be done with an easy addition or subtraction.

Example: $123 - 96$
Solution by steps:
(*a*) Add 4 to 96 to make it an easy 100
(*b*) $123 - 100 = 23$; this is 4 less than the real difference
(*c*) Add the 4 back: $23 + 4 = 27$, answer

Or you can add the same amount to both numbers, like this:
(*a*) $123 + 4 = 127$
(*b*) $96 + 4 = 100$,
(*c*) $127 - 100 = 27$

Short way: Add 4 to both numbers: $127 - 100 = 27$

Why it works: In subtraction, if you add the same amount to both numbers, you do not change the difference between the numbers. Try it with any numbers, such as $5 - 1 = 4$.

Now add 2 to both numbers: $(5 + 2) - (1 + 2) = 7 - 3 = 4$.

Or add 3 to both numbers: $(5 + 3) - (1 + 3) = 8 - 4 = 4$.

Instead of adding, you may find it easier to subtract the same amount from both numbers. Again, this will not change the difference between the numbers.

Example: $224 - 173$
Solution: Subtract 3 from both numbers; this makes 173 end in zero; then: $221 - 170 = 51$, answer

3-3. Subtraction without Borrowing

You can sometimes avoid borrowing if you subtract by pairs of digits instead of by individual digits.

Example (1):
$$\begin{array}{r} 5,381 \\ -\ 4,863 \\ \hline 518 \end{array}$$

Solution: Instead of borrowing 1 from 8 and then subtracting 3 from 11, you can do this subtraction in two easy steps:
(*a*) $81 - 63 = 18$
(*b*) $53 - 48 = 5$; answer 518

Example (2):
$$\begin{array}{r} 47,635 \\ -\ 38,519 \\ \hline 9,116 \end{array}$$

Solution: (*a*) $35 - 19 = 16$
(*b*) $6 - 5 = 1$
(*c*) $47 - 38 = 9$; answer 9,116

3-4. Splitting Numbers into Easy Parts

You can split numbers into smaller parts for easier subtraction.

Example: $3,862 - 2,428$

Solution by steps:
(*a*) Split 3,862 into $3,800 + 62$
(*b*) Split 2,428 into $\underline{2,400} + \underline{28}$
(*c*) Subtract: $1,400 + 34 = 1,434$, answer

Or you can do it this way:
$3,800 - 2,400 = 1,400; 62 - 28 = 34; 1,400 + 34 = 1,434$

3-5. Counting Change after a Purchase

If you buy some groceries for $3.69 and pay for them with a $5 bill, how much change should you get?

Perhaps you can subtract $3.69 from $5.00 mentally in one step and see the correct remainder as $1.31. If so, fine! If not, you may find it easier to calculate the cents and the dollars separately. Most of the numbers you use will end in zero or zeros so that they are easy to handle.

Solution by steps:

(a) Add 1¢ to the 69¢; = 70¢. Add 30¢ to reach the next dollar ($4.00). So your change in coins is 31¢.

(b) Subtract $4.00 from $5.00; = $1.00, your change in dollars

(c) Total change: 31¢ + $1.00 = $1.31

Or you can check your change *after* you get it by using the money as "counters." Start with the purchase amount and add the coins up to the next dollar; then add the dollars in change until you reach the denomination of the bill you gave in payment.

3-6. Checking Your Answer in Subtraction

The fastest, easiest way to prove your subtraction is to add the remainder to the number that was subtracted. If the sum of these two numbers equals the number from which the subtraction was made, your answer may be considered correct.

Or, to put it another way, you add the remainder to the number above it to see if their sum equals the number on top.

Example: 634,715
          − 491,284
           143,431 Remainder

Check: Adding 143,431 to 491,284 gives 634,715

Since you don't have to write any additional numbers to make this check, the method is fast and easy, and should be used to prove every subtraction. Many errors in calculations are the result of errors in subtraction so that taking the time to check remainders will not only promote greater accuracy but will save time and effort in the long run.

Of course, if you want to be even more certain of your answer, you can also subtract the remainder from the number on top to see if you get the number that was subtracted. Again, this can be done without writing any additional numbers. In the example given, you would subtract 143,431 from 634,715 to get 491,284, indicating that the answer is correct.

You can also check subtraction by casting out 9's but the method is not included here because the check by adding the remainder to the number subtracted is faster, easier, and surer.

IV. MULTIPLICATION

4-1. Basic Principles

Multiplication is already a fast process. It is a rapid method for adding the same number over and over.

Example:
```
        513              513
        513            ×   4
        513            2,052
        513
      2,052
```

Multiplication may be written in one of the following ways:

$3 \times 8 = 24$ Product

```
  3  Factor (multiplicand)
× 8  Factor (multiplier)
 24  Product
```

The numbers 3 and 8 are called *factors* of 24 because, when multiplied together, they give 24. Other factors of 24 are 2, 4, 6, and 12 since $2 \times 12 = 24$, $4 \times 6 = 24$, $2 \times 2 \times 2 \times 3 = 24$, etc.

Besides being a *product* of 3×8, 4×6, etc., 24 is also said to be a *multiple* of 2, 3, 4, 6, 8, and 12 because each of these factors can be multiplied by another factor or factors to produce 24.

Numbers may be multiplied in any order. Thus, $3 \times 8 = 24$; $8 \times 3 = 24$; $3 \times 8 = 8 \times 3$. You can take advantage of this fact by multiplying in whatever order is easiest for you.

Example: $25 \times 42 \times 4$
Solution: $25 \times 4 = 100$; $\times 42 = 4{,}200$, answer

Another basic way to simplify multiplication is to split a number into smaller parts by units, tens, hundreds, etc.

Example: 47×6

Solution by steps:

(*a*) Think of 47 as $40 + 7$

(*b*) Multiply by parts: $40 \times 6 = 240$; $7 \times 6 = 42$

(*c*) Add the partial products: $240 + 42 = 282$, answer

Or you can think of it this way:

$$40 \times 6 = 240$$
$$\underline{7 \times 6 = 42}$$
$$47 \times 6 = 282$$

Or do it this way: $(40 \times 6) + (7 \times 6) = 240 + 42 = 282$

To simplify multiplication, see if you can change a multiplier to one of the easiest numbers: 0, 1, 10 (also 100, 1,000, etc.), or 2. The products of these multipliers are as follows:

Any number multiplied by zero gives zero.
Examples: $1 \times 0 = 0$; $2 \times 0 = 0$; $3 \times 0 = 0$

Any number multiplied by 1 gives the same number.
Examples: $2 \times 1 = 2$; $3 \times 1 = 3$; $4 \times 1 = 4$

Any number multiplied by 10 gives the same number with a zero added or the decimal point moved one place to the right.
Examples: $36 \times 10 = 360$; $4.3 \times 10 = 43$; $.62 \times 10 = 6.2$

Any number multiplied by 2 equals double the number.
Examples: $24 \times 2 = 48$; $45 \times 2 = 90$

Numbers ending in 5 (15, 25, 35, etc.) and numbers near a 10-level (9, 11, 19, 21, etc.) can be made into easy multipliers by transforming them into numbers ending in zero.

Example (1): 34×15
Solution by steps:
(*a*) Double $15 = 30$
(*b*) Halve $34 = 17$
(*c*) $17 \times 30 = 510$, answer

Example (2): 27×9
(*a*) Change 9 to a multiplier made up of two easy numbers: $10 - 1$
(*b*) $27 \times 10 = 270$; $- (27 \times 1) = 243$, answer

Applications and further details of these short-cut principles will be found in the methods that follow.

4-2. Multiplication When One Number Has All Digits Alike

If the digits of a number are all the same, make this number the multiplier.

Example: 88 × 76
Solution: Make 88 the multiplier. This will require multiplying only by 8 instead of by 7 and 6.

```
      76
  ×  88
     608
    608
   6688
```

With larger numbers, the method becomes even faster and easier.

Example: 888 × 764

```
       764                    888
   ×  888                 ×  764
      6112                   3552
      6112                   5328
     6112                    6216
    678432                  678432
```

4-3. To Multiply by a Number Made Up of Multiples

Example (1): 753 × 842
In the multiplier, 4 is a multiple of 2 since 2 × 2 = 4; 8 is a multiple of 4 since 4 × 2 = 8.
Solution:

```
      753
  ×  842
    1506    Multiply this by 2 instead of 753 by 4
   3012     Multiply this by 2 instead of 753 by 8
  6024
 634026
```

Example (2): 243 × 186
In this multiplier, 18 is a multiple of 6 since 6 × 3 = 18.
Solution:

```
      243
  ×  186
    1458
   4374    Obtained by multiplying 1458 by 3 since 18 is 3 times 6.
  45198    This is instead of multiplying 243 by 8 and then by 1.
```

Proving the answer:

$$\begin{array}{r} 243 \\ \times\ 186 \\ \hline 1458 \\ 1944 \\ 243\ \ \\ \hline 45198 \end{array}$$

4-4. No-Carry Multiplication

If carrying is something you would like to avoid, you can get around it this way:

$$\begin{array}{r} 634 \\ \times\ 9 \\ \hline 36 \\ 27 \\ 54\ \ \\ \hline 5706 \end{array}$$

You can also do no-carry multiplication by splitting a number into units, tens, hundreds, etc., and multiplying from left to right. The zeros in the partial products help to make the addition easier.

$$\begin{array}{r} 634 \\ \times\ \ 9 \\ \hline 5400 \\ 270 \\ 36 \\ \hline 5706 \end{array}$$

Also: $\begin{aligned} 600 \times 9 &= 5,400 \\ 30 \times 9 &= 270 \\ 4 \times 9 &= 36 \\ \hline 634 \times 9 &= 5,706 \end{aligned}$

Or you can do it this way:

$(600 \times 9) + (30 \times 9) + (4 \times 9) = 5,400 + 270 + 36 = 5,706$

4-5. To Multiply a Number by 10, 100, 1,000, etc.

Multiplying by 10, 100, etc., is so simple that it hardly need be considered as a separate step. However, the process is reviewed here because the use of these multipliers is very important to methods that come later.

A. If the number has a decimal point:

Move the point one place to the right for each zero in the multiplier.

To multiply by 10, move the decimal point one place to the right; to multiply by 100, move it two places to the right; to multiply by 1,000, move the decimal point three places to the right, etc.

Examples: $6.27 \times 10 = 62.7$; $.183 \times 10 = 1.83$
$6.27 \times 100 = 627$; $.183 \times 100 = 18.3$
$6.27 \times 1,000 = 6,270$; $.183 \times 1,000 = 183$

B. If the number has no decimal point:
Add a zero after the number for each zero in the multiplier.
To multiply by 10, add one zero after the number; to multiply by 100, add two zeros; to multiply by 1,000, add three zeros, etc.

Examples: $73 \times 10 = 730$; $192 \times 10 = 1,920$
$73 \times 100 = 7,300$; $192 \times 100 = 19,200$
$73 \times 1,000 = 73,000$; $192 \times 1,000 = 192,000$

In the following methods, where a number is to be multiplied by 10, 100, or 1,000, simply move the decimal point to the right as required, or add the proper number of zeros.

Multiplying by Aliquot Parts of 1, 10, 100, and 1,000

An aliquot part is a number that is contained in another number an exact number of times. For example, 5 is an aliquot part of 10 because 5 can be divided into 10 without a remainder.

Aliquot parts of 1, 10, 100, and 1,000 (such as .5, 5, 2.5, 25, 50, 125, etc.) can be simplified as multipliers by changing them into easy fractions, as shown in the following methods.

4-6. General Method for Multiplying by Aliquot Parts

To multiply by an aliquot part, change it to its fractional equivalent and multiply. The numerator (top) of the fraction will be 1, 10, 100, or 1,000 while the denominator of the fraction will be 2, 4, or 8. By this method, .5 is changed to the fraction $\frac{1}{2}$, 5 is changed to $\frac{10}{2}$, 25 to $\frac{100}{4}$, 125 to $\frac{1000}{8}$, etc.

Example (1): 16×25
Solution by steps:
(a) Change 25 to fractional equivalent: $\frac{100}{4}$

(b) $16 \times \frac{100}{4} = \frac{1600}{4} = 400$, answer

Short way: $16 \div 4 = 4$; add two zeros $= 400$
Proving the answer: 16
                              $\times \underline{25}$
                               80
                              $\underline{32}$
                              400

Example (2):   16 × 2.5
Solution by steps:
(*a*) Fractional equivalent of 2.5 = $\frac{10}{4}$
(*b*) 16 × $\frac{10}{4}$ = $\frac{160}{4}$ = 40, answer

Short way: 16 ÷ 4 = 4; add a zero = 40
Why it works: The multipliers (2.5, 25, etc.) are not altered in value but are simply changed to an easier form. This does not change the result or product.

Although 1 is included among the multipliers (along with 10, 100, and 1,000) in the methods that follow, actually multiplying by 1 should not be done since it will not change the number multiplied. Also keep in mind that multiplying by $\frac{1}{2}$, $\frac{1}{4}$, or $\frac{1}{8}$ is the same as dividing by 2, 4, or 8. (For details of multiplying and dividing by fractions, see Chapter VI.)

4-7. To Multiply by .5, 5, 50, or 500

The general method for this group of multipliers is to divide by 2 and then multiply by 10, 100, or 1,000. (No need to multiply by 1 as it will not change the number multiplied.)

A. To multiply a number by .5, divide the number by 2. Dividing by 2 is the same as multiplying by $\frac{1}{2}$ which is equivalent to .5 ($\frac{5}{10}$).

Example: 362 × .5
Solution: 362 ÷ 2 = 181;   or,   362 × $\frac{1}{2}$ = 181,   answer

Proving the answer:
$$\begin{array}{r} 362 \\ \times\quad .5 \\ \hline 181.0 \end{array}$$

B. To multiply a number by 5, divide the number by 2 and multiply by 10. This is the same as multiplying by $\frac{10}{2}$ which is equivalent to 5.
Example: 482 × 5
Solution: 482 ÷ 2 = 241;   × 10 = 2,410, answer
Proving the answer:
$$\begin{array}{r} 482 \\ \times\quad 5 \\ \hline 2,410 \end{array}$$

C. To multiply a number by 50, divide the number by 2 and multiply by 100. This is the same as multiplying by $\frac{100}{2}$ which is equivalent to 50.

Example: 138 × 50
Solution: 138 ÷ 2 = 69; × 100 = 6,900, answer
Proving the answer: 138
× 50
6,900

D. To multiply a number by 500, divide the number by 2 and multiply by 1,000. This is the same as multiplying by $\frac{1,000}{2}$ which is equivalent to 500.

Example: 45.62 × 500
Solution: 45.62 ÷ 2 = 22.81; × 1,000 = 22,810, answer

Proving the answer: 45.62   Short way: 4562
× 500   × 5
22810.00   22810

E. If the number to be multiplied is odd, there will be a remainder of $\frac{1}{2}$ after dividing by 2 but the same steps apply.

Examples: 47 × 5: 47 ÷ 2 = $23\frac{1}{2}$ = 23.5; × 10 = 235
23 × 50: 23 ÷ 2 = $11\frac{1}{2}$ = 11.5; × 100 = 1,150
85 × 500: 85 ÷ 2 = $42\frac{1}{2}$ = 42.5; × 1,000 = 42,500

However, you can multiply odd numbers by these multipliers in a still shorter way: simply ignore the fraction ($\frac{1}{2}$) after dividing by 2 and attach the multiplier (.5, 5, 50, or 500) to the result.

Examples: 23 × .5 = 11.5 (halve 22 and attach .5)
69 × 5 = 345 (halve 68 and attach 5)
57 × 50 = 2,850 (halve 56 and attach 50)
73 × 500 = 36,500 (halve 72 and attach 500)

4-8. To Multiply by .25, 2.5, 25, or 250

The general method for this group of multipliers is to divide by 4 and then multiply by 10, 100, or 1,000. (No need to multiply by 1 as it will not change the number multiplied.)

A. To multiply a number by .25, divide the number by 4. Dividing by 4 is the same as multiplying by $\frac{1}{4}$ which is equivalent to .25 ($\frac{25}{100}$).

Example: 72 × .25

Solution: 72 ÷ 4 = 18; or, 72 × $\frac{1}{4}$ = 18, answer

Multiplication ■ **35**

Proving the answer:

$$\begin{array}{r} 72 \\ \times .25 \\ \hline 360 \\ 144 \\ \hline 18.00 \end{array}$$

B. To multiply a number by 2.5, divide the number by 4 and multiply by 10. This is the same as multiplying by $\frac{10}{4}$ which is equivalent to 2.5 $(2\frac{1}{2})$.

Example: 96 × 2.5
Solution: 96 ÷ 4 = 24; × 10 = 240, answer

Proving the answer:

$$\begin{array}{r} 96 \\ \times 2.5 \\ \hline 480 \\ 192 \\ \hline 240.0 \end{array}$$

C. To multiply a number by 25, divide the number by 4 and multiply by 100. This is the same as multiplying by $\frac{100}{4}$ which is equivalent to 25.

Example: 428 × 25
Solution: 428 ÷ 4 = 107; × 100 = 10,700, answer

Proving the answer:

$$\begin{array}{r} 428 \\ \times 25 \\ \hline 2140 \\ 856 \\ \hline 10700 \end{array}$$

D. To multiply a number by 250, divide the number by 4 and multiply by 1,000. This is the same as multiplying by $\frac{1000}{4}$ which is equivalent to 250.

Example: 848 × 250
Solution: 848 ÷ 4 = 212; × 1,000 = 212,000, answer

Proving the answer:

$$\begin{array}{r} 848 \\ \times 250 \\ \hline 42400 \\ 1696 \\ \hline 212000 \end{array}$$

E. If the number to be multiplied is not evenly divisible by 4, there will be a remainder of $\frac{1}{4}$ (.25), $\frac{1}{2}$ (.5), or $\frac{3}{4}$ (.75) but the same steps apply.

Examples: $17 \times .25$: $17 \div 4 = 4\frac{1}{4}$ or 4.25
               18×2.5: $18 \div 4 = 4\frac{1}{2} = 4.5$; $\times 10 = 45$
               19×25: $19 \div 4 = 4\frac{3}{4} = 4.75$; $\times 100 = 475$
             21×250: $21 \div 4 = 5\frac{1}{4} = 5.25$; $\times 1{,}000 = 5{,}250$

4-9. To Multiply by .125, 1.25, 12.5, or 125

The general method for this group of multipliers is to divide by 8 and then multiply by 10, 100, or 1,000. (No need to multiply by 1 as it will not change the number multiplied.)

A. To multiply a number by .125, divide the number by 8. Dividing by 8 is the same as multiplying by $\frac{1}{8}$ which is equivalent to .125 $(\frac{125}{1000})$.

Example: $168 \times .125$
Solution: $168 \div 8 = 21$;   or, $168 \times \frac{1}{8} = 21$, answer

Proving the answer:
$$
\begin{array}{r}
168 \\
\times\,.125 \\
\hline
840 \\
336 \\
168 \\
\hline
21.000
\end{array}
$$

B. To multiply a number by 1.25, divide the number by 8 and multiply by 10. This is the same as multiplying by $\frac{10}{8}$ which is equivalent to 1.25 $(1\frac{1}{4})$.

Example: 328×1.25
Solution: $328 \div 8 = 41$; $\times 10 = 410$, answer

Proving the answer:
$$
\begin{array}{r}
328 \\
\times\,1.25 \\
\hline
1640 \\
656 \\
328 \\
\hline
410.00
\end{array}
$$

C. To multiply a number by 12.5, divide the number by 8 and multiply by 100. This is the same as multiplying by $\frac{100}{8}$ which is equivalent to 12.5 $(12\frac{1}{2})$.

Example: 48 × 12.5
Solution: 48 ÷ 8 = 6; × 100 = 600, answer

Proving the answer:

$$\begin{array}{r} 48 \\ \times\ 12.5 \\ \hline 240 \\ 96 \\ 48 \\ \hline 600.0 \end{array}$$

D. To multiply a number by 125, divide the number by 8 and multiply by 1,000. This is the same as multiplying by $\frac{1000}{8}$ which is equivalent to 125.

Example: 728 × 125
Solution: 728 ÷ 8 = 91; × 1,000 = 91,000, answer

Proving the answer:

$$\begin{array}{r} 728 \\ \times\ 125 \\ \hline 3640 \\ 1456 \\ 728 \\ \hline 91000 \end{array}$$

E. *Alternate methods* of multiplying by .125, 1.25, 12.5, and 125.

1. To multiply a number by .125, instead of dividing by 8, divide by 2 and then 4, or divide by 2, then 2, and by 2 again.

   Example: 432 × .125: 432 ÷ 2 = 216; ÷ 2 = 108; ÷ 2 = 54, answer

2. To multiply a number by 1.25, add $\frac{1}{4}$ of the number to the number.

   Example: 168 × 1.25 = 168 + 42 = 210, answer

3. To multiply a number by 12.5, multiply by 10 and add $\frac{1}{4}$ of the product to the product.

   Example: 48 × 12.5 = 480 + 120 = 600, answer

4. To multiply a number by 125, multiply by 100 and add $\frac{1}{4}$ of the product to the product.

   Example: 728 × 125 = 72,800 + 18,200 = 91,000, answer

4-10. To Multiply by .75, 7.5, 75, or 750

While .75, 7.5, 75, and 750 are not evenly divisible into 1, 10, 100, or 1,000 and are therefore not aliquot parts of these numbers, they are included here because they are closely related to the aliquot parts just discussed and can be converted into easy multipliers by similar steps.

The general method for this group of multipliers is to multiply by $\frac{3}{4}$ and then by 10, 100, or 1,000. (No need to multiply by 1 as it will not change the number multiplied.)

A. To multiply a number by .75, multiply the number by $\frac{3}{4}$ which is equivalent to .75 ($\frac{75}{100}$).

Example: 64 × .75
Solution: 64 × $\frac{3}{4}$ = 48; or, 64 ÷ 4 = 16; × 3 = 48, answer

Proving the answer:
$$\begin{array}{r} 64 \\ \times\ .75 \\ \hline 320 \\ 448 \\ \hline 48.00 \end{array}$$

B. To multiply a number by 7.5, divide the number by 4 and multiply by 30 (or by 3 and then by 10). This is the same as multiplying by $\frac{30}{4}$ which is equivalent to 7.5 ($7\frac{1}{2}$).

Example: 92 × 7.5
Solution: 92 ÷ 4 = 23; × 3 = 69; × 10 = 690, answer

Proving the answer:
$$\begin{array}{r} 92 \\ \times\ 7.5 \\ \hline 460 \\ 644 \\ \hline 690.0 \end{array}$$

C. To multiply a number by 75, divide the number by 4 and multiply by 300 (or by 3 and then by 100). This is the same as multiplying by $\frac{300}{4}$ which is equivalent to 75.

Example: 48 × 75
Solution: 48 ÷ 4 = 12; × 3 = 36; × 100 = 3,600, answer

Proving the answer:
$$\begin{array}{r} 48 \\ \times\ 75 \\ \hline 240 \\ 336 \\ \hline 3600 \end{array}$$

Multiplication ▪ 39

D. To multiply a number by 750, divide the number by 4 and multiply by 3,000 (or by 3 and then by 1,000). This is the same as multiplying by $\frac{3000}{4}$ which is equivalent to 750.

Example: 648 × 750

Solution: 648 ÷ 4 = 162; × 3 = 486; × 1,000 = 486,000, answer

Proving the answer:

$$\begin{array}{r} 648 \\ \times\ \underline{750} \\ 32400 \\ \underline{4536\ \ } \\ 486000 \end{array}$$

E. If the number to be multiplied is not evenly divisible by 4, there will be a remainder of $\frac{1}{4}$ (.25), $\frac{1}{2}$ (.5), or $\frac{3}{4}$ (.75) but the same steps apply.

Examples:
21 × .75: 21 ÷ 4 = $5\frac{1}{4}$; × 3 = $15\frac{3}{4}$ or 15.75
22 × 7.5: 22 ÷ 4 = $5\frac{1}{2}$; × 3 = $16\frac{1}{2}$ = 16.5; × 10 = 165
23 × 75: 23 ÷ 4 = $5\frac{3}{4}$; × 3 = $17\frac{1}{4}$ = 17.25: × 100 = 1,725
26 × 750: 26 ÷ 4 = $6\frac{1}{2}$; × 3 = $19\frac{1}{2}$ = 19.5; × 1,000 = 19,500

F. *Alternate method* of multiplying by .75, 7.5, 75, and 750.

1. To multiply a number by .75, halve the number and to this add $\frac{1}{2}$ of the half.

    Example: 64 × .75 = 32 + 16 = 48, answer

2. To multiply a number by 7.5, halve the number, to this add $\frac{1}{2}$ of the half, and multiply by 10.

    Example: 92 × 7.5: 46 + 23 = 69; × 10 = 690, answer

3. To multiply a number by 75, halve the number, to this add $\frac{1}{2}$ of the half, and multiply by 100.

    Example: 48 × 75: 24 + 12 = 36; × 100 = 3,600, answer

4. To multiply a number by 750, halve the number, to this add $\frac{1}{2}$ of the half, and multiply by 1,000.

    Example: 648 × 750: 324 + 162 = 486; × 1,000 = 486,000, answer

G. *A third method* of multiplying by .75, 7.5, 75, and 750.

1. To multiply a number by .75, subtract $\frac{1}{4}$ of the number.

   Example: $64 \times .75 = 64 - 16 = 48$, answer

2. To multiply a number by 7.5, subtract $\frac{1}{4}$ of the number and multiply by 10.

   Example: 92×7.5: $92 - 23 = 69$; $\times 10 = 690$, answer

3. To multiply a number by 75, subtract $\frac{1}{4}$ of the number and multiply by 100.

   Example: 48×75: $48 - 12 = 36$; $\times 100 = 3,600$, answer

4. To multiply a number by 750, subtract $\frac{1}{4}$ of the number and multiply by 1,000.

   Example: 648×750: $648 - 162 = 486$; $\times 1,000$
   $= 486,000$, answer

Summary of Multiplication by Aliquot Parts and 75-Numbers

N stands for the number to be multiplied

To Multiply by	Method
.5	N/2 (N ÷ 2)
5	N/2 × 10
50	N/2 × 100
500	N/2 × 1,000
.25	N/4
2.5	N/4 × 10
25	N/4 × 100
250	N/4 × 1,000
.125	N/8
1.25	N/8 × 10
12.5	N/8 × 100
125	N/8 × 1,000
.75	N/4 × 3
7.5	N/4 × 3 × 10
75	N/4 × 3 × 100
750	N/4 × 3 × 1,000

Other aliquot parts: Multiplication by other aliquot parts in the form of mixed numbers (whole numbers with fractions, such as $12\frac{1}{2}$, $16\frac{2}{3}$, and $33\frac{1}{3}$) is discussed in Chapter VI.

To Obtain the Squares of Certain Numbers

As you probably know, the "square" of a number is simply

the product obtained by multiplying the number by itself. Thus, 16 is the square of 4 since $4 \times 4 = 16$.

4-11. To Square a Number Ending in 1

Example: 41×41

Solution by steps:

(a) Square the next lower number (always ends in zero): $40 \times 40 = 1,600$

(b) Obtain the sum of the next lower number and the number being squared: $40 + 41 = 81$

(c) Add (a) and (b): $1,600 + 81 = 1,681$, answer

Short way: $40 \times 40 = 1,600$; $40 + 41 = 81$; answer $1,681$

Proving the answer:
$$
\begin{array}{r}
41 \\
\times\ \underline{41} \\
41 \\
\underline{164} \\
1681
\end{array}
$$

Other examples: $21 \times 21 = 441$
$31 \times 31 = 961$
$61 \times 61 = 3,721$

4-12. To Square a Number Ending in 4

Example: 34×34

Solution by steps:

(a) Square the next higher number: $35 \times 35 = 1,225$ (see Method 4-13)

(b) Obtain the sum of the number being squared and the next higher number: $34 + 35 = 69$

(c) Subtract (b) from (a): $1,225 - 69 = 1,156$, answer (simplify your subtraction by adding 1 to both numbers: $1,226 - 70 = 1,156$)

Short way: $35 \times 35 = 1,225$; $-(34 + 35) = 1,156$

Proving the answer:
$$
\begin{array}{r}
34 \\
\times\ \underline{34} \\
136 \\
\underline{102} \\
1156
\end{array}
$$

Why it works: How the square of a number equals the square of the next higher number minus the sum of the higher number and the number being squared is shown in the following:

5×5: $(6 \times 6) - (6 + 5) = 36 - 11 = 25$
4×4: $(5 \times 5) - (5 + 4) = 25 - 9 = 16$
3×3: $(4 \times 4) - (4 + 3) = 16 - 7 = 9$
2×2: $(3 \times 3) - (3 + 2) = 9 - 5 = 4$
1×1: $(2 \times 2) - (2 + 1) = 4 - 3 = 1$

4-13. To Square a Number Ending in 5
Example: 35×35

Solution by steps:
(a) Multiply the first digit by the digit plus 1: $3 \times 4 = 12$
(b) Attach 25: 1,225, answer

Short way: $3 \times 4 = 12$; answer 1,225

Proving the answer:
$$\begin{array}{r} 35 \\ \times\ 35 \\ \hline 175 \\ 105 \\ \hline 1225 \end{array}$$

Other examples: $4.5 \times 4.5 = 20.25$
$65 \times 65 = 4,225$
$.75 \times .75 = .5625$
$115 \times 115 = 13,225$ (treat 11 like a "first digit")

4-14. To Square a Number Ending in 6
This works by the same steps as for numbers ending in 1 (Method 4-11).
Example: 36×36

Solution by steps:
(a) Square the next lower number: $35 \times 35 = 1,225$ (by Method 4-13)

(b) Obtain the sum of the next lower number and the number being squared: $35 + 36 = 71$

(c) Add (a) and (b): $1,225 + 71 = 1,296$, answer

Short way: $35 \times 35 = 1,225$; $+ (35 + 36) = 1,296$

Multiplication ▪ **43**

Proving the answer:

$$\begin{array}{r} 36 \\ \times\ \underline{36} \\ 216 \\ \underline{108} \\ 1296 \end{array}$$

Why it works: How the square of a number equals the square of the next lower number plus the sum of the lower number and the number being squared is shown in the following:

$$2 \times 2:\ (1 \times 1) + (1 + 2) = 1 + 3 = 4$$
$$3 \times 3:\ (2 \times 2) + (2 + 3) = 4 + 5 = 9$$
$$4 \times 4:\ (3 \times 3) + (3 + 4) = 9 + 7 = 16$$
$$5 \times 5:\ (4 \times 4) + (4 + 5) = 16 + 9 = 25$$
$$6 \times 6:\ (5 \times 5) + (5 + 6) = 25 + 11 = 36$$

4-15. To Square a Number Ending in 9

This works by the same steps as for numbers ending in 4 (Method 4-12).

Example: 39×39

Solution by steps:

(*a*) Square the next higher number: $40 \times 40 = 1,600$

(*b*) Obtain the sum of the number being squared and the next higher number: $39 + 40 = 79$

(*c*) Subtract (*b*) from (*a*): $1,600 - 79 = 1,521$, answer

Short way: $40 \times 40 = 1,600$; $-(39 + 40) = 1,521$

Proving the answer:

$$\begin{array}{r} 39 \\ \times\ \underline{39} \\ 351 \\ \underline{117} \\ 1521 \end{array}$$

Other examples:

$$19 \times 19 = 361$$
$$49 \times 49 = 2,401$$
$$79 \times 79 = 6,241$$

Summary of Squaring Certain Numbers

N stands for the Number to be squared

Number ends in 1 or 6: $(N - 1) \times (N - 1) + (N + N - 1)$

Number ends in 4 or 9: $(N + 1) \times (N + 1) - (N + N + 1)$
Number ends in 5 (two-digit N): 1st digit $\times$ (1st digit + 1);
    attach 25
Number ends in 5 (three-digit N): 1st two digits $\times$
    (1st two digits + 1); attach 25

Multiplying Numbers That Have Small Differences

4-16. To Multiply Two Numbers with a Difference of 1

Square either number, whichever is easier. If you square the larger number, subtract it from its square; if you square the smaller number, add it to its square.

Example (1):   24 $\times$ 23

Solution by steps:
(a) Square the larger number; this is easy to do by Method 4-12: 24 $\times$ 24 = 576

(b) Subtract the larger number: 576 – 24 = 552, answer
Short way: 24 $\times$ 24 = 576; – 24 = 552

Proving the answer:
$$\begin{array}{r} 24 \\ \times\ 23 \\ \hline 72 \\ 48 \\ \hline 552 \end{array}$$

Why it works: Squaring a number and subtracting the number from its square is equivalent to multiplying the number by the number minus 1.

Example (2):   35 $\times$ 36

Solution by steps:
(a) Square the smaller number; this is easy to do by Method 4-13: 35 $\times$ 35 = 1,225

(b) Add the smaller number: 1,225 + 35 = 1,260, answer

Short way: 35 $\times$ 35 = 1,225; + 35 = 1,260

Proving the answer:
$$\begin{array}{r} 35 \\ \times\ 36 \\ \hline 210 \\ 105 \\ \hline 1260 \end{array}$$

Why it works: Squaring a number and adding the number to its square is equivalent to multiplying the number by the number plus 1.

Other examples:
Squaring the higher number: $40 \times 39 = 1,600 - 40 = 1,560$
Squaring the lower number: $40 \times 41 = 1,600 + 40 = 1,640$

4-17. To Multiply Two Numbers with a Difference of 2
Square the average of the two numbers and subtract 1.
Example: 19×21

Solution by steps:

(*a*) The average is the middle or "in-between" number: 20

(*b*) Square the average: $20 \times 20 = 400$

(*c*) Subtract 1: $400 - 1 = 399$, answer

Short way: $20 \times 20 = 400$; $-1 = 399$

Proving the answer:
$$\begin{array}{r} 19 \\ \times\ 21 \\ \hline 19 \\ 38 \\ \hline 399 \end{array}$$

Why it works: If you multiply two small numbers with a difference of 2, you can see how the method works. For example, 8×6. The average of these numbers is 7 and the problem could be written as $(7 + 1) \times (7 - 1)$. By cross multiplying, here is how it works out:

$$\begin{array}{r} 7 + 1 \\ \times\ 7 - 1 \\ \hline 49 + 7 \\ -\ 7 - 1 \\ \hline 49\ \ \ \ -1 \end{array}$$

The $+7$ and -7 cancel each other, leaving $49 - 1$ which equals 48, the product of 8×6. Thus, the square of the average (7×7) minus 1 equals 8×6.

4-18. To Multiply Two Numbers with a Difference of 3
Example: 27×24

Solution by steps:
(*a*) Add 1 to the smaller number and square the sum:
$24 + 1 = 25$; $\times 25 = 625$ (by Method 4-13)

(*b*) Take 1 from the smaller number and add to (*a*):
24 – 1 = 23; + 625 = 648, answer

Short way: 25 × 25 = 625; + 23 = 648

Proving the answer: 27
                        × 24
                        ────
                        108
                         54
                        ────
                        648

Why it works: If you multiply two small numbers with a difference of 3, you can see how the method works. For example, 8 × 11. (8 + 1) = 9; × 9 = 81; (8 – 1) = 7; + 81 = 88

Other examples: 92 × 89 = 8,100 + 88 = 8,188
                   57 × 54 = 3,025 + 53 = 3,078

4-19. To Multiply Two Numbers with a Difference of 4
Square the average of the two numbers and subtract 4.
Example: 67 × 63

Solution by steps:
(*a*) Average: 65 (67 –2 or 63 + 2)

(*b*) 65 × 65 = 4,225 (by Method 4-13)

(*c*) 4,225 – 4 = 4,221, answer

Short way: 65 × 65 = 4,225; – 4 = 4,221

Proving the answer: 67
                        × 63
                        ────
                        201
                        402
                        ────
                        4221

Why it works: If you multiply two small numbers with a difference of 4, you can see how the method works.

Example: 5 × 9; average, 7; × 7 = 49; – 4 = 45
           Or, consider 5 as (7 – 2) and 9 as (7 + 2);
           (7 – 2) × (7 + 2) = 49 – 4 = 45

Other examples: 78 × 82 = 6,400 – 4 = 6,396
                   63 × 59 = 3,721 – 4 = 3,717
                   (61 × 61 = 3,721 by Method 4-11)

Multiplication ▪ **47**

4-20. To Multiply Two Numbers with a Difference of 6

Square the average of the two numbers and subtract 9.
Example: 34 × 28

Solution by steps:

(*a*) Average: 31 (34 – 3 or 28 + 3)

(*b*) 31 × 31 = 961 (by Method 4-11)

(*c*) 961 – 9 = 952, answer

Short way: 31 × 31 = 961; – 9 = 952

Why it works: If you multiply two small numbers with a difference of 6, you can see how the method works.

Example: 2 × 8; average, 5; × 5 = 25; – 9 = 16
Or, consider 2 as (5 – 3) and 8 as (5 + 3);
(5 – 3) × (5 + 3) = 25 – 9 = 16

Other examples: 83 × 77 = 6,400 – 9 = 6,391
48 × 42 = 2,025 – 9 = 2,016

Summary of Multiplying Numbers Having Small Differences

The following table is based on 10 and numbers near 10

Difference of 1: 10 × 9 = 90 = (10 × 10) –10 = 90
Difference of 1: 10 × 9 = 90 = (9 × 9) + 9 = 90
Difference of 2: 11 × 9 = 99 = (10 × 10) – 1 = 99
Difference of 3: 11 × 8 = 88 = (8 + 1) × (8 + 1)
+ (8 – 1) = (9 × 9) + 7 = 88
Difference of 4: 12 × 8 = 96 = (10 × 10) – 4 = 96
Difference of 6: 13 × 7 = 91 = (10 × 10) – 9 = 91

Multiplying by Numbers Ending in 1

4-21. General Method for Multiplying by Numbers Ending in 1

Since 11 = 10 + 1, 21 = 20 + 1, etc., you can multiply by 11, 21, etc., more easily if you multiply by 10, 20, etc., and then add the number multiplied. The number you add is the "+1" of 10 + 1, 20 + 1, and so on.
Example: 67 × 11

Solution: 67 × (10 + 1) = 670 + 67 = 737, answer

Other examples: 36 × 21 = 720 + 36 = 756
27 × 31 = 810 + 27 = 837
49 × 101 = 4,900 + 49 = 4,949

Further details and alternate short-cuts for multiplying by 1-numbers will be found in the following methods.

4-22. To Multiply a Number by 11

Method A: This is the method explained in 4-21. Multiply the number by 10 and add the number.

Method B: Write the number to be multiplied and add 10 times the number.

Example: 369 × 11

Solution:

	369	Also:		369
	3690			369
	4059			4059

Method C: To multiply a two-digit number by 11, add the number's digits and insert their sum between the digits.

Example: 27 × 11

Solution by steps:
(a) Sum of the number's digits: $2 + 7 = 9$
(b) Insert the 9 between the 2 and 7: 297, answer

Short way: $2 + 7 = 9$; answer 297

Other examples: $43 \times 11 = 473$
$72 \times 11 = 792$
$85 \times 11 = 935$ ($8 + 5 = 13$; place 3 in the middle; carry 1 to 8; becomes 9)

4-23. To Multiply a Number by 21

Double the number, multiply this by 10, and add the number.

Example: 67 × 21

Solution: $2 \times 67 = 134$; $\times 10 = 1,340$; $+ 67 = 1,407$, answer

Proving the answer:

$$\begin{array}{r} 67 \\ \times\ \ 21 \\ \hline 67 \\ 134\ \ \\ \hline 1407 \end{array}$$

Multiplication ■ **49**

4-24. To Multiply a Number by 31, 41, 51, etc.

Multiply the number by the first digit of the multiplier, multiply the result by 10, and add the number.

Example: 24 × 31

Solution by steps:
(a) Multiply 24 by the first digit of 31: 24 × 3 = 72
(b) 72 × 10 = 720; + 24 = 744, answer

Short way: 24 × 3 = 72; × 10 = 720; + 24 = 744

Proving the answer:

$$\begin{array}{r} 24 \\ \times\ 31 \\ \hline 24 \\ 72\ \ \\ \hline 744 \end{array}$$

Multiplying by Numbers Ending in 5

4-25. To Multiply by 15, 25, 35, 45, etc.
Halve-and-Double Method

Halve the number to be multiplied and double the multiplier. This changes the multiplier ending in 5 to a multiplier ending in zero.

Example: 28 × 15

Solution: Half of 28 = 14; double 15 = 30; × 14 = 420, answer

Proving the answer:

$$\begin{array}{r} 28 \\ \times\ 15 \\ \hline 140 \\ 28\ \ \\ \hline 420 \end{array}$$

Why it works: In multiplication, halving one number (dividing it by 2) and doubling the other number (multiplying it by 2) does not change the product. This is easy to see in an example using small numbers, such as 4 × 6 = 24. Double 4 = 8; halve 6 = 3; 8 × 3 = 24. Or, halve 4 = 2; double 6 = 12; 2 × 12 = 24, again.

Other examples: 64 × 35 = 32 × 70 = 2,240
22 × 45 = 11 × 90 = 990
48 × 3.5 = 24 × 7 = 168

If the number to be multiplied is odd, the method still applies.
Example: 17 × 45

Solution: Half of 17 = $8\frac{1}{2}$;   double 45 = 90;   × $8\frac{1}{2}$ = 765, answer

Sometimes you can halve-and-double twice to simplify the multiplication further.
Example: 175 × 24

Solution: 350 × 12 = 700 × 6 = 4,200, answer

4-26. Alternate Method for Multiplying a Number by 15

Add half the number to itself and multiply by 10.
Example (1):   64 × 15

Solution: 64 + 32 = 96;   × 10 = 960, answer

Proving the answer:
$$\begin{array}{r} 64 \\ \times\ 15 \\ \hline 320 \\ 64 \\ \hline 960 \end{array}$$

Example (2):   65 × 15

Solution: 65 + $32\frac{1}{2}$ = $97\frac{1}{2}$ = 97.5;   × 10 = 975, answer

Why it works: Adding half the number to itself is the same as multiplying the number by $1\frac{1}{2}$; then multiplying this by 10 is the same as multiplying the original number by 15.

4-27. Alternate Method for Multiplying a Number by 45

Multiply the number by 50 and subtract $\frac{1}{10}$ of the product.
Example: 36 × 45

Solution by steps:
(a) 36 × 50 = 1,800 (36 ÷ 2 = 18;   × 100 = 1,800 by Method 4-7C)

(b) $\frac{1}{10}$ × 1,800 = 180

(c) 1,800 – 180 = 1,620, answer

Short way: 36 × 50 = 1,800;   – 180 = 1,620

Why it works: Multiplying by 50 and subtracting $\frac{1}{10}$ of the product is the same as multiplying by $(50 - 5)$ which is equivalent to 45.

4-28. Alternate Method for Multiplying a Number by 55

Multiply the number by 50 and add $\frac{1}{10}$ of the product.
Example: 48×55

Solution by steps:
(*a*) $48 \times 50 = 2,400$ ($48 \div 2 = 24$; $\times 100 = 2,400$)

(*b*) $\frac{1}{10} \times 2,400 = 240$

(*c*) $2,400 + 240 = 2,640$, answer

Short way: $48 \times 50 = 2,400$; $+ 240 = 2,640$

Why it works: Multiplying by 50 and adding $\frac{1}{10}$ of the product is the same as multiplying by $(50 + 5)$ which is equivalent to 55.

4-29. To Multiply One Number by Another When They Both End in 5 and the Sum of Their Other Digits is Even

Example: 35×55

Solution by steps:
(*a*) Multiply the first digits: $3 \times 5 = 15$

(*b*) Obtain half the sum of the first digits:
$3 + 5 = 8$; $\times \frac{1}{2} = 4$

(*c*) Add (*a*) and (*b*): $15 + 4 = 19$

(*d*) Attach 25 to (*c*): 1,925, answer

Short way: $3 \times 5 = 15$; $+ \frac{8}{2} = 19$; answer 1,925

Proving the answer:

$$
\begin{array}{r}
35 \\
\times 55 \\
\hline
175 \\
175 \\
\hline
1925
\end{array}
$$

Why it works:
(*a*) Multiplying 3 by 5 is really $30 \times 50 = 1,500$

(*b*) Adding 4 to 15 is really adding 400 to $1,500 = 1,900$

(*c*) The 25 attached to 19 is the product of 5×5: answer 1,925

Other examples: $85 \times 45 = 3,825$
                       $4.5 \times 6.5 = 29.25$
                       $75 \times .35 = 26.25$
                       $125 \times 85 = 10,625$

4-30. To Multiply One Number by Another When They Both End in 5 and the Sum of Their Other Digits Is Odd

This follows the same steps as Method 4-29. However, there is a fraction, which is dropped, and the answer ends in 75 instead of 25.

Example: 65×35
Solution by steps:

(a) Multiply the first digits: $6 \times 3 = 18$

(b) Obtain half the sum of the first digits:
     $6 + 3 = 9$; $\times \frac{1}{2} = 4\frac{1}{2}$;   drop the $\frac{1}{2}$ to get 4

(c) Add (a) and (b): $18 + 4 = 22$

(d) Attach 75 to (c): 2,275, answer

Proving the answer:
$$\begin{array}{r} 65 \\ \times\ 35 \\ \hline 325 \\ 195 \\ \hline 2275 \end{array}$$

Why it works:

(a) Multiplying 6 by 3 is really $60 \times 30 = 1,800$

(b) The $4\frac{1}{2}$ is really 450; dropping the $\frac{1}{2}$ is dropping 50

(c) Adding 4 to 18 is really adding 400 to 1,800: $= 2,200$; this is 50 too little because of the 50 dropped in (b)

(d) The 75 attached to 22 (really added to 2,200) is the product of 5×5 plus the 50 dropped in (b): answer 2,275

Other examples: $75 \times 45 = 3,375$
                       $3.5 \times 6.5 = 22.75$
                       $.95 \times 85 = 80.75$
                       $115 \times 125 = 14,375$

Multiplying by 9, by Numbers Ending in 9, and by Multiples of 9

The general method for multiplying by these numbers is to increase them to the next number ending in zero; then, after multiplying, to subtract the excess amount from the product, as shown in the following methods.

4-31. To Multiply a Number by 9

Multiply the number by 10 and subtract the number.

Example: 37×9

Solution: $37 \times 10 = 370$; $-37 = 333$, answer

Proving the answer:
$$\begin{array}{r} 37 \\ \times\ \underline{\ 9} \\ 333 \end{array}$$

Why it works: Multiplying a number by 10 and subtracting the number is the same as multiplying by (10 – 1) which is equivalent to 9. You get the same result if you subtract $\frac{1}{10}$ of the product after multiplying the number by 10.

4-32. To Multiply a Number by 19, 29, 39, etc.

Add 1 to the multiplier (19, 29, etc.) to make it end in zero; then multiply the given number, and subtract the number from the result.

Example: 24×19

Solution: $19 + 1 = 20$; $\times 24 = 480$; $-24 = 456$, answer

Proving the answer:
$$\begin{array}{r} 24 \\ \times\ \underline{19} \\ 216 \\ \underline{24\ \ } \\ 456 \end{array}$$

Why it works: Multiplying a number by 20 and subtracting the number is the same as multiplying by (20 – 1) which is equivalent to 19. Other numbers ending in 9 work the same way: 29 = (30 – 1), 39 = (40 – 1), etc.

Other examples: $36 \times 29 = 1{,}044$
$48 \times 59 = 2{,}832$
$2.1 \times 69 = 144.9$

4-33. To Multiply by a Multiple of 9 from 18 to 81

This refers to multiplying by 18, 27, 36, 45, 54, 63, 72, and 81 which are multiples of 9 produced by 9×2, 9×3, 9×4, etc.

Example: 35 × 27

Solution by steps:

(*a*) Increase the multiple of 9 to the next number ending in zero. Thus, increase 27 to 30

(*b*) Multiply: 35 × 30 = 1,050

(*c*) Subtract $\frac{1}{10}$ of the product: 1,050 – 105 = 945, answer

Short way: 35 × 30 = 1,050; – 105 = 945

Proving the answer:
```
        35
      × 27
       245
        70
       945
```

Why it works: Multiplying a number by 30 and subtracting $\frac{1}{10}$ of the product is the same as multiplying the number by (30 –3) which is equivalent to 27. The other multiples of 9 work the same way: 18 = 20 – 2; 36 = 40 – 4; 45 = 50 – 5; 54 = 60 – 6, etc.

Other examples: 65 × 18 = 1,170
92 × 36 = 3,312
2.4 × 6.3 = 15.12

4-34. To Multiply a Number by 99

Multiply the number by 100 and subtract the number.

Example: 83 × 99

Solution: 83 × 100 = 8,300; – 83 = 8,217

Why it works: Multiplying a number by 100 and subtracting the number is the same as multiplying the number by (100 – 1) which is equivalent to 99.

Other examples: 46 × 99 = 4,554
3.7 × 99 = 366.3
248 × 99 = 24,552

4-35. To Multiply by a Multiple of 99

This refers to multiplying by 198, 297, 396, 495, 594, 693, 792, 891, and 990 which are multiples of 99 produced by 99 × 2, 99 × 3, 99 × 4, etc.

Example: 13 × 297

Solution by steps:

(a) Increase the multiple of 99 to the next hundred. Thus, increase 297 to 300

(b) Multiply: $13 \times 300 = 3,900$

(c) Subtract $\frac{1}{100}$ of the product: $3,900 - 39 = 3,861$, answer

Proving the answer:

$$\begin{array}{r} 297 \\ \times\ \ 13 \\ \hline 891 \\ 297\ \ \ \\ \hline 3861 \end{array}$$

Why it works: Multiplying a number by 300 and subtracting $\frac{1}{100}$ of the product is the same as multiplying the number by $(300 - 3)$ which is equivalent to 297. The other multiples of 99 work the same way: $198 = 200 - 2$; $396 = 400 - 4$; $495 = 500 - 5$, etc.

Other examples:

$$24 \times 198 = 4,752$$
$$62 \times 396 = 24,552$$
$$.71 \times 4.95 = 3.5145$$

Multiplying by Numbers Near 100

The general method for simplifying multiplication by numbers near 100 is to change the number into an easy multiplier consisting of 100 and a small number, as follows: change 98 to $100 - 2$; 99 to $100 - 1$; 101 to $100 + 1$; 102 to $100 + 2$, etc.

4-36. To Multiply a Number by 98

Multiply the number by 100 and subtract twice the number.

Example: 64×98

Solution: $64 \times 100 = 6,400$; $64 \times 2 = 128$;
$6,400 - 128 = 6,272$, answer

Proving the answer:

$$\begin{array}{r} 64 \\ \times\ \ 98 \\ \hline 512 \\ 576\ \ \ \\ \hline 6272 \end{array}$$

Why it works: Multiplying a number by 100 and subtracting twice the number is the same as multiplying the number by

(100 – 2) which is equivalent to 98. Other numbers near 100 work the same way.

To Multiply a Number by 99:
Multiply the number by 100 and subtract the number, as previously described in Method 4-34.

4-37. To Multiply a Number by 101
Multiply the number by 100 and add the number.

Example: 64 × 101

Solution: 64 × 100 = 6,400; + 64 = 6,464, answer

To multiply any number from 10 to 99 by 101, simply write the number twice.

Examples: 24 × 101 = 2,424
76 × 101 = 7,676
93 × 101 = 9,393

4-38. To Multiply a Number by 102
Multiply the number by 100 and add twice the number.
Example: 46 × 102

Solution: 46 × 100 = 4,600; 46 × 2 = 92;
+ 4,600 = 4,692, answer

Proving the answer:

$$\begin{array}{r} 102 \\ \times\ \underline{46} \\ 612 \\ \underline{408\ } \\ 4692 \end{array}$$

Other Short-Cuts in Multiplication

4-39. To Multiply One "Teen" Number by Another
Actually, this method can be used with 11 and 12 as well as the "teen" numbers from 13 to 19.
Example: 16 × 13

Solution by steps:
(*a*) Add either number to the units digit of the other:
16 + 3 (or 13 + 6) = 19

(*b*) Multiply by 10: 19 × 10 = 190

(*c*) Multiply the units digits: $6 \times 3 = 18$

(*d*) Add (*b*) and (*c*): $190 + 18 = 208$, answer

Short way: $16 + 3 = 19$; $\times 10 = 190$; $+ (6 \times 3) = 208$

Proving the answer:

$$\begin{array}{r} 16 \\ \times\ 13 \\ \hline 48 \\ 16 \\ \hline 208 \end{array}$$

Other examples: $14 \times 19 = 266$
$18 \times 17 = 306$
$15 \times 13 = 195$

If the last two digits add to 10, simply obtain the product of the last two digits and place a 2 in front of it.

Example: 17×13

Solution: $7 \times 3 = 21$; answer 221

Other examples: $19 \times 11 = 209$ (zero must be inserted)
$18 \times 12 = 216$
$16 \times 14 = 224$
$15 \times 15 = 225$ (also by Method 4-13)

The next method extends this idea to use with other numbers and shows that the 2 in the above examples is derived by multiplying $1 \times (1 + 1)$, or 1×2.

4-40. To Multiply Two Numbers When Their End Digits Add to 10 and Their Other Digits Are the Same

Example: 74×76

Solution by steps:
(*a*) Multiply the first digit by the first digit $+1$: $7 \times 8 = 56$

(*b*) Multiply the end digits: $4 \times 6 = 24$

(*c*) Combine (*a*) and (*b*): 5,624, answer

Short way: $7 \times 8 = 56$; $4 \times 6 = 24$; answer 5,624

Proving the answer:

$$\begin{array}{r} 74 \\ \times\ 76 \\ \hline 444 \\ 518 \\ \hline 5624 \end{array}$$

Why it works: Multiplying 7 by 8 is really $70 \times 80 = 5,600$. Why multiply by 80? Because there are seventy 70's plus ten 70's from multiplying by 6 and by 4. The 24 in the answer is, of course, the product of 4×6.

You can get a better look at how this works by using no-carry multiplication (Method 4-4), as follows:

$$
\begin{array}{r}
74 \\
\times \ \ 76 \\
\hline
4900 \\
280 \\
420 \\
24 \\
\hline
5624
\end{array}
\begin{array}{l}
(70 \times 70) \\
(70 \times \ 4) \\
(6 \times 70) \\
(6 \times 4)
\end{array}
$$

$$280 \ (70 \times 4) \atop 420 \ (6 \times 70) \Big\} = 10 \times 70 = 700$$

This can be shortened to:

$$
\begin{array}{l}
70 \times 70 = 4900 \\
10 \times 70 = \ \ 700
\end{array} \Big\} = 80 \times 70 = 5,600
$$
$$
\underline{6 \times \ 4 = \ \ \ 24} \\
 5624
$$

And shortened again to: $(80 \times 70) + (6 \times 4) = 5,600 + 24 = 5,624$

And finally to: $8 \times 7 = 56$; attach 24; $= 5,624$

Each set of numbers in the teens, 20's, 30's, 40's, etc., contains five pairs of numbers that can be multiplied by this method. For example, in the 60's:

$$
\begin{array}{l}
69 \times 61 = 4,209 \ \text{(zero must be inserted)} \\
68 \times 62 = 4,216 \\
67 \times 63 = 4,221 \\
66 \times 64 = 4,224 \\
65 \times 65 = 4,225 \ \text{(same as by Method 4-13)}
\end{array}
$$

These pairs, of course, can also be reversed since $69 \times 61 = 61 \times 69$, $68 \times 62 = 62 \times 68$, etc.

The idea can be extended to larger numbers too, with the first two digits on the left treated as a "first digit."

Example: 113×117

Solution: $11 \times (11 + 1) = 11 \times 12 = 132$; $3 \times 7 = 21$; answer 13,221

Other examples: $254 \times 256 = 65,024$
$418 \times 412 = 172,216$

4-41. To Multiply Two Two-Digit Numbers When Their First Digits Add to 10 and Their End Digits Are the Same

Example: 36 × 76

Solution by steps:
(a) Multiply the first digits and add an end digit:
   3 × 7 = 21; + 6 = 27
(b) Multiply the end digits: 6 × 6 = 36
(c) Combine (a) and (b): 2,736, answer

Short way: (3 × 7) + 6 = 27; 6 × 6 = 36; answer 2,736

Proving the answer:
$$\begin{array}{r} 36 \\ \times\ \underline{76} \\ 216 \\ \underline{252} \\ 2736 \end{array}$$

Further examples are given in the following complete set of numbers in the 20's paired with numbers in the 80's having the same end digits. Notice that a zero must be inserted in the tens place of the first four products.

20 × 80 = 1,600	25 × 85 = 2,125
21 × 81 = 1,701	26 × 86 = 2,236
22 × 82 = 1,804	27 × 87 = 2,349
23 × 83 = 1,909	28 × 88 = 2,464
24 × 84 = 2,016	29 × 89 = 2,581

Starting with 10 × 90, 11 × 91, 12 × 92, etc., you can multiply 50 pairs of two-digit numbers by this method. These pairs can also be reversed since 10 × 90 = 90 × 10, etc.

Checking Results in Multiplication

4-42. Estimate the Answer Before You Multiply

Estimating your answer before you multiply will help to avoid big mistakes, such as getting an answer in the hundreds when it should be in the thousands, or the other way around, as may happen if you misplace a decimal point or put the right digit in the wrong place. If your answer comes out very different in magnitude from your estimate, check to see what is wrong.

One way to estimate the answer is to round off the numbers for an easy multiplication. Change the numbers in opposite directions, that is, increase one number and decrease the other. This will produce a smaller change in the product than if you increase or decrease both numbers.

For example, you might estimate the product of 486 × 112 like this:

Rounded off to the nearest hundred: 500 × 100 = 50,000
Rounded off to the nearest ten: 490 × 110 = 53,900

You can also estimate the answer by a short-cut method, such as by multiplying 486 by 100 to get 48,600 and adding $\frac{1}{8}$ of 48,600, since 112 is approximately $\frac{1}{8}$ larger than 100. This would give you an estimate of 54,675.

Any of these estimates will serve to show you in advance that your answer should be somewhere around 50,000. After you do the multiplication, check your estimate against your answer.

$$\begin{array}{r} 486 \\ \times\ 112 \\ \hline 972 \\ 486 \\ 486 \\ \hline 54432 \end{array}$$

Now you can use one of the following methods to see if your multiplication is correct.

4-43. Checking Multiplication by Multiplying in Reverse

To make this check, reverse the numbers and multiply again.

Example:
$$\begin{array}{r} 49 \\ \times\ 36 \\ \hline 294 \\ 147 \\ \hline 1764 \end{array}$$

Check:
$$\begin{array}{r} 36 \\ \times\ 49 \\ \hline 324 \\ 144 \\ \hline 1764 \end{array}$$

4-44. Checking Multiplication by Using a Short-Cut

See if you can use a short-cut for a fast, easy check on your answer. For example, you could check the product (1,764) obtained in Method 4-43 as follows:

49 × 36 = (50 × 36) – 36 = 1,800 – 36 = 1,764 (by Method 4-32)

If you use a short-cut in the first place, you can check your answer by another short-cut or by conventional multiplication.

4-45. Checking Multiplication by Dividing the Product by One of the Factors Used

Dividing the product by one of the factors should give the other factor as the quotient. This check takes longer but you might want to make it if the result is of great importance. Divide by the multiplier (rather than by the number multiplied) since this will change the order in which the digits are multiplied.

Example:
$$\begin{array}{r} 374 \text{ Factor} \\ \times \quad 85 \text{ Factor} \\ \hline 1870 \\ 2992 \\ \hline 31790 \end{array}$$

Check:
$$\begin{array}{r} 374 \text{ Quotient} \\ 85\overline{)31790} \\ 255 \\ \hline 629 \\ 595 \\ \hline 340 \\ 340 \end{array}$$

4-46. Checking Multiplication by Casting Out 9's

Various ways to cast out 9's are shown in 1-6, Chapter I. The application of this method to checking multiplication is discussed here. Use of the method is recommended in particular when large numbers are involved. One of its advantages is that you do not have to write the numbers again.

The remainders obtained after casting out 9's are used as check numbers for testing the answer in multiplication, as follows:

Example (1):
$$\begin{array}{r} 752 \text{ Digit sum: 14; Remainder: 5} \\ \times \quad 436 \text{ Digit sum: 13; Remainder: 4} \\ \hline 4512 \quad 5 \times 4 = 20; \text{ Remainder: 2} \\ 2256 \\ 3008 \\ \hline 327872 \text{ Digit sum: 29; Remainder: 2} \end{array}$$

Multiplying the remainders of the factors gives a remainder of 2. The remainder of the answer is also 2. Therefore, the answer checks and may be considered correct.

In the following example, no digit sums are shown; the 9's are cast out by omitting them.

Example (2):   639 Remainder: 0
                 × 453 Remainder: 3
                  1917   0 × 3 = 0 Remainder
                  3195
                  2556
                289467 Remainder: 0

Since the remainders are the same (0), the answer checks as correct.

If the remainder of either factor is zero, the product of the remainders of the factors will also be zero since zero multiplied by any number is always zero.

Remember that casting out 9's does not give positive proof of the answer. The method will not catch errors of the following types:

1. Where digits have been transposed, such as in 3,451 and 3,541. Since the same digits appear in both numbers, their digit sums and remainders will naturally be the same.

2. Where the decimal point has been misplaced. Again, it is obvious that 3.451, 34.51, 345.1, and 3,451 will all give the same digit sums and remainders. (You can guard against errors of this type by estimating your answer in advance.)

However, if casting out 9's does *not* confirm your answer, you can be sure that there is something wrong in the multiplication, in the check, or in both.

V. DIVISION

5-1. Basic Principles

Division is the process of finding how many times one number is contained in another. It is a rapid way of subtracting. Thus, $8 \div 2 = 4$ means that 2 is contained in 8 four times, or that 2 can be subtracted from 8 four times.

Division may also be shown in the form of a fraction:
$$8 \div 2 = \tfrac{8}{2} = 4.$$

Division is the inverse of multiplication. If a product is divided by one of two factors, the result (quotient) will be the other factor.

Multiplication	*Division*

$$
\begin{array}{rl}
154 & \text{Factor} \\
\times\underline{6} & \text{Factor} \\
924 & \text{Product}
\end{array}
$$

$$
\text{Divisor } 6\overline{)\,924\ \text{Dividend}} \\
\phantom{\text{Divisor } 6)\,}154\ \text{Quotient}
$$

$$
\phantom{\text{Divisor } 154)\,}6\ \text{Quotient} \\
\text{Divisor } 154\overline{)\,924\ \text{Dividend}} \\
\phantom{\text{Divisor } 154)\,}\underline{924}
$$

One basic way to simplify division is to break up the divisor into easy factors. You can then divide by these smaller numbers in whatever order is easiest for you. To divide 378 by 18, for example, you can factor 18 into 6×3 and divide in either order: $378 \div 6 = 63$; $\div 3 = 21$; or $378 \div 3 = 126$; $\div 6 = 21$. Or you can use the factors 2×9: $378 \div 2 = 189$; $\div 9 = 21$.

Another basic way of simplifying division depends on the fact that the divisor and the dividend can be multiplied by the same number, or be divided by the same number, without changing the answer (quotient). For example, to divide 385 by 35, you can double both numbers to get $770 \div 70$ which, of course, is an easy 11.

These and other ways of making division faster and easier are covered in detail under the methods that follow.

5-2. Doing Short Division

In "short" division, the multiplying and dividing steps are not written out. This method is generally employed for dividing by numbers up to 12 although you can use it with larger divisors, also. Two simple procedures can help to make your short division faster, easier, and surer.

A. Write each digit of your answer directly under the proper digit in the dividend.

B. Write each partial remainder as a small number in front of and above (or below) the next digit in the dividend. This will save you from having to remember each remainder and will make it easier to check your division in case of a mistake.

Example: Divisor 6) 1⁵7⁰4,⁹9⁷7²2 Dividend
                        29,162 Quotient

5-3. Doing Long Division

In "long" division, the multiplying and subtracting steps are written out so that the remainders are shown for each step. The following procedures can help you get the right answer faster and easier.

A. Write each digit of your answer directly above the proper digit in the dividend.

B. When a digit repeats in the answer, use the product previously obtained with this digit instead of multiplying again.

Example:

```
        717
432) 309744
    3024      The first product of 7 × 432, 3024, should
     734      simply be written for the second 7 instead of
     432      multiplying by 7 again.
    3024
    3024
```

C. When a digit in the answer is a multiple of a preceding digit in the answer, you can simplify your multiplication as shown below.

Example:

```
       36
48) 1728      Instead of multiplying 48 by 6, multiply
    144       144 by 2 since 6 is twice 3. Thus, 6 × 48
    288       = 288; 2 × 144 = 288
    288
```

D. If the divisor ends in one or more zeros, cancel these before you divide; move the decimal point in the dividend one place to the left for each zero cancelled.

Example: 173,216 ÷ 3,200

$$
\begin{array}{r}
54.13 \\
32\cancel{00}\overline{)\,1732.16} \\
\underline{160} \\
132 \\
\underline{128} \\
41 \\
\underline{32} \\
96 \\
\underline{96}
\end{array}
$$

Why it works: Cancelling the two zeros and moving the decimal point two places to the left is the same as dividing both numbers by 100. In division, the answer is not changed if the divisor and the dividend are both divided by the same number (or multiplied by the same number).

If you do the division WITHOUT cancelling the zeros, you have to handle much larger numbers.

$$
\begin{array}{r}
54.13 \\
3200\overline{)\,173216.00} \\
\underline{16000} \\
13216 \\
\underline{12800} \\
4160 \\
\underline{3200} \\
9600 \\
\underline{9600}
\end{array}
$$

5-4. To Divide by 10, 100, 1,000, etc.

Dividing by 10, 100, etc., is so simple that it hardly need be considered as a separate step. However, the process is reviewed here because the use of these divisors is very important to methods that come later.

A. To divide a number by 10, move the decimal point one place to the left.

Examples: 682 ÷ 10 = 68.2   72.31 ÷ 10 = 7.231

B. To divide a number by 100, move the decimal point two places to the left.

Examples: 361 ÷ 100 = 3.61   4.26 ÷ 100 = .0426

C. To divide a number by 1,000, move the decimal point three places to the left.

Examples: $945 \div 1{,}000 = .945$ $80 \div 1{,}000 = .08$

In the following methods, where a number is to be divided by 10, 100, or 1,000, simply move the decimal point the required number of places to the left.

Dividing by Aliquot Parts of 1, 10, 100, and 1,000

As mentioned in Chapter IV on multiplying, an aliquot part is a number that goes into another number an exact number of times. For example, 5 is an aliquot part of 10 because 5 can be divided into 10 "evenly", that is, without a remainder.

Aliquot parts of 1, 10, 100, and 1,000 (such as .5, 5, .25, 2.5, 12.5, and 125) can be simplified as divisors by changing them into easy fractions, as shown in the following methods.

5-5. General Method for Dividing by Aliquot Parts

Change the aliquot part to its fractional equivalent, invert the fraction, and multiply.

Example: $641 \div 25$

Solution by steps:

(*a*) 25 is an aliquot part of 100. Change it to its fractional equivalent: $\frac{100}{4}$

(*b*) Invert the fraction: $\frac{4}{100}$ (equivalent to $\frac{1}{25}$)

(*c*) Multiply: $641 \times \frac{4}{100} = \frac{2564}{100} = 25.64$, answer

Short way: $641 \times 4 = 2{,}564;$ answer 25.64

Why it works: Dividing by 25 is the same as multiplying by $\frac{1}{25}$ which is equivalent to $\frac{4}{100}$. Multiplying by 4 and pointing off two decimal places to the left is the same as multiplying by $\frac{4}{100}$.

Dividing by other aliquot parts works on the same principle. The divisors discussed in the next methods, with their fractional equivalents, are:

$.5 = \frac{1}{2};$ $5 = \frac{10}{2};$ $50 = \frac{100}{2};$ $500 = \frac{1000}{2}$

$.25 = \frac{1}{4};$ $2.5 = \frac{10}{4};$ $25 = \frac{100}{4};$ $250 = \frac{1000}{4}$

$.125 = \frac{1}{8};$ $1.25 = \frac{10}{8};$ $12.5 = \frac{100}{8};$ $125 = \frac{1000}{8}$

For multiplying and dividing by fractions, see Chapter VI.

5-6. To Divide by .5, 5, 50, or 500

A. To divide a number by .5, multiply the number by 2. Multiplying by 2 is the same as dividing by $\frac{1}{2}$ which is equivalent to .5 $(\frac{5}{10})$.

Example: $85 \div .5$

Solution: $85 \times 2 = 170$, answer

Proving the answer: $.5\overline{)85} = 5\overline{)850}$
$$170$$

B. To divide a number by 5, multiply the number by 2 and divide by 10. This is the same as multiplying by $\frac{2}{10}$ which is equivalent to dividing by $\frac{10}{2}$ which equals 5.

Example: $243 \div 5$

Solution: $243 \times 2 = 486$; $\div 10 = 48.6$, answer

Proving the answer: $5\overline{)243.0}$
$$48.6$$

C. To divide a number by 50, multiply the number by 2 and divide by 100. This is the same as multiplying by $\frac{2}{100}$ which is equivalent to dividing by $\frac{100}{2}$ which equals 50.

Example: $215 \div 50$

Solution: $215 \times 2 = 430$; $\div 100 = 4.3$, answer

Proving the answer:

$$50\overline{)215.0} \quad \text{Short way:} \quad 5\overline{)21.5}$$

with intermediate steps:
$$\begin{array}{r} 4.3 \\ 50\overline{)215.0} \\ 200 \\ \hline 150 \\ 150 \\ \hline \end{array} \qquad \begin{array}{r} 5\overline{)21.5} \\ 4.3 \end{array}$$

D. To divide a number by 500, multiply the number by 2 and divide by 1,000. This is the same as multiplying by $\frac{2}{1000}$ which is equivalent to dividing by $\frac{1000}{2}$ which equals 500.

Example: $164 \div 500$

Solution: $164 \times 2 = 328$; $\div 1,000 = .328$, answer

Proving the answer:

$$\begin{array}{r} .328 \\ 500\overline{)164.000} \\ 1500 \\ \hline 1400 \\ 1000 \\ \hline 4000 \\ 4000 \\ \hline \end{array} \qquad \text{Short way:} \quad \begin{array}{r} 5\overline{)1.640} \\ .328 \end{array}$$

5-7. To Divide by .25, 2.5, 25, or 250

A. To divide a number by .25, multiply the number by 4. Multiplying by 4 is the same as dividing by $\frac{1}{4}$ which is equivalent to .25 $\left(\frac{25}{100}\right)$.

Example: $71 \div .25$

Solution: $71 \times 4 = 284$, answer

Proving the answer:

$$.25 \overline{)71} = 25 \overline{)\begin{array}{r} 284 \\ 7100 \end{array}}$$

$$\begin{array}{r} \underline{50} \\ 210 \\ \underline{200} \\ 100 \\ \underline{100} \end{array}$$

B. To divide a number by 2.5, multiply the number by 4 and divide by 10. This is the same as multiplying by $\frac{4}{10}$ which is equivalent to dividing by $\frac{10}{4}$ which equals 2.5 $(2\frac{1}{2})$.

Example: $52 \div 2.5$

Solution: $52 \times 4 = 208$; $\div 10 = 20.8$, answer

Proving the answer:

$$2.5 \overline{)52} = 25 \overline{)\begin{array}{r} 20.8 \\ 520.0 \end{array}}$$

$$\begin{array}{r} \underline{50} \\ 200 \\ \underline{200} \end{array}$$

C. To divide a number by 25, multiply the number by 4 and divide by 100. This is the same as multiplying by $\frac{4}{100}$ which is equivalent to dividing by $\frac{100}{4}$ which equals 25.

Example: $161 \div 25$

Solution: $161 \times 4 = 644$; $\div 100 = 6.44$, answer

Proving the answer:

$$25 \overline{)\begin{array}{r} 6.44 \\ 161.00 \end{array}}$$

$$\begin{array}{r} \underline{150} \\ 110 \\ \underline{100} \\ 100 \\ \underline{100} \end{array}$$

D. To divide a number by 250, multiply the number by 4 and divide by 1,000. This is the same as multiplying by $\frac{4}{1000}$ which is equivalent to dividing by $\frac{1000}{4}$ which equals 250.

Example: 832 ÷ 250

Solution: 832 × 4 = 3,328; ÷ 1,000 = 3.328, answer

Proving the answer:

$$
\begin{array}{r}
3.328 \\
250\overline{)832.000} \\
\underline{750} \\
820 \\
\underline{750} \\
700 \\
\underline{500} \\
2000 \\
\underline{2000}
\end{array}
\quad \text{or:} \quad
\begin{array}{r}
3.328 \\
25\overline{)83.200} \\
\underline{75} \\
82 \\
\underline{75} \\
70 \\
\underline{50} \\
200 \\
\underline{200}
\end{array}
$$

5-8. To Divide by .125, 1.25, 12.5, or 125

A. To divide a number by .125, multiply the number by 8. This is the same as dividing by $\frac{1}{8}$ which is equivalent to .125 ($\frac{125}{1000}$).

Example: 32 ÷ .125

Solution: 32 × 8 = 256, answer

Proving the answer:

$$
.125\overline{)32} = 125\overline{)32000}
\begin{array}{r}
256 \\
\underline{250} \\
700 \\
\underline{625} \\
750 \\
\underline{750}
\end{array}
$$

B. To divide a number by 1.25, multiply the number by 8 and divide by 10. This is the same as multiplying by $\frac{8}{10}$ which is equivalent to dividing by $\frac{10}{8}$ which equals 1.25 ($1\frac{1}{4}$ or $\frac{5}{4}$).

Example: 53 ÷ 1.25

Solution: 53 × 8 = 424; ÷ 10 = 42.4, answer

Proving the answer:

$$
1.25\overline{)53} = 125\overline{)5300.0}
\begin{array}{r}
42.4 \\
\underline{500} \\
300 \\
\underline{250} \\
500 \\
\underline{500}
\end{array}
$$

C. To divide a number by 12.5, multiply the number by 8 and divide by 100. This is the same as multiplying by $\frac{8}{100}$ which is equivalent to dividing by $\frac{100}{8}$ which equals 12.5 ($12\frac{1}{2}$ or $\frac{25}{2}$).

Example: 315 ÷ 12.5

Solution: 315 × 8 = 2,520; ÷ 100 = 25.2, answer

Proving the answer:

$$12.5\overline{)315} = 125\overline{)3150.0}$$

with the long division showing:

```
            25.2
125)3150.0
    250
    650
    625
    250
    250
```

D. To divide a number by 125, multiply the number by 8 and divide by 1,000. This is the same as multiplying by $\frac{8}{1000}$ which is equivalent to dividing by $\frac{1000}{8}$ which equals 125.

Example: 441 ÷ 125

Solution: 441 × 8 = 3,528; ÷ 1,000 = 3.528, answer

Proving the answer:

```
            3.528
125)441.000
    375
    660
    625
    350
    250
    1000
    1000
```

5-9. To Divide by .75, 7.5, 75, or 750

While .75, 7.5, 75, and 750 are not evenly divisible into 1, 10, 100, or 1,000 and are therefore not aliquot parts of these numbers, they are included here because they are closely related to the aliquot parts just discussed and can be converted into easy divisors by similar steps.

The fractional equivalents of these divisors are: $.75 = \frac{3}{4}$; $7.5 = \frac{30}{4}$; $75 = \frac{300}{4}$; and $750 = \frac{3000}{4}$. When these fractions are inverted and used as multipliers, this involves multiplying by 4, dividing by 3, and then dividing by 10, 100, or 1,000, as the case may be. (No need to multiply by 1 as it will not change the number multiplied.) Since multiplying by 4 and dividing by 3 is the same as multiplying by $\frac{4}{3}$, you may find it easier to simply increase the multiplied number by $\frac{1}{3}$ of the number. This is the same as multiplying by $1\frac{1}{3}$ which equals $\frac{4}{3}$.

A. To divide a number by .75, increase the number by $\frac{1}{3}$ of the number.

Example: $36 \div .75$

Solution: $36 \times \frac{1}{3} = 12$; $+ 36 = 48$, answer

also: $36 \times \frac{4}{3} = 48$

Proving the answer:

$$.75\overline{)36} = 75\overline{)3600} \;\; \overset{48}{}$$

$$\underline{300}$$
$$600$$
$$\underline{600}$$

B. To divide a number by 7.5, increase the number by $\frac{1}{3}$ of the number and divide by 10.

Example: $144 \div 7.5$

Solution: $144 \times \frac{1}{3} = 48$; $+ 144 = 192$; $\div 10 = 19.2$, answer

also: $144 \times \frac{4}{3} = 192$; $\div 10 = 19.2$

Proving the answer:

$$7.5\overline{)144} = 75\overline{)1440.0} \;\; \overset{19.2}{}$$

$$\underline{75}$$
$$690$$
$$\underline{675}$$
$$150$$
$$\underline{150}$$

C. To divide a number by 75, increase the number by $\frac{1}{3}$ of the number and divide by 100.

Example: $825 \div 75$

Solution: $825 \times \frac{1}{3} = 275$; $+ 825 = 1,100$; $\div 100 = 11$, answer

also: $825 \times \frac{4}{3} = 1,100$; $\div 100 = 11$

Proving the answer:

$$75\overline{)825} \;\; \overset{11}{}$$

$$\underline{75}$$
$$75$$
$$\underline{75}$$

D. To divide a number by 750, increase the number by $\frac{1}{3}$ of the number and divide by 1,000.

Example: 126 ÷ 750

Solution: 126 × $\frac{1}{3}$ = 42; + 126 = 168; ÷ 1,000 = .168, answer

also: 126 × $\frac{4}{3}$ = 168; ÷ 1,000 = .168

Proving the answer:

```
        .168              .168
750) 126.000   or:  75) 12.600
     750                75
     5100               510
     4500               450
     6000               600
     6000               600
```

Summary of Division by Aliquot Parts and 75-Numbers

N stands for the Number to be divided

To Divide by	Method
.5	2N (2 × N)
5	2N ÷ 10
50	2N ÷ 100
500	2N ÷ 1,000
.25	4N
2.5	4N ÷ 10
25	4N ÷ 100
250	4N ÷ 1,000
.125	8N
1.25	8N ÷ 10
12.5	8N ÷ 100
125	8N ÷ 1,000
.75	4N/3 (or N + N/3)
7.5	4N/3 ÷ 10
75	4N/3 ÷ 100
750	4N/3 ÷ 1,000

Other aliquot parts: Division by other aliquot parts in the form of mixed numbers ($2\frac{1}{2}$, $12\frac{1}{2}$, $16\frac{2}{3}$, etc.) is discussed in Chapter VI.

Testing the Divisibility of a Number

It is often useful to be able to tell quickly if one number can be divided evenly by another. The following tests make this easy to do.

5-10. Test for Divisibility by Odd and Even Numbers

A. *Odd numbers:* Numbers ending in 1, 3, 5, 7, or 9 can only be divided evenly by another odd number. Therefore, if a number's last digit is odd, you know at once that the number is not evenly divisible by any even number.

B. *Even numbers:* Numbers ending in 2, 4, 6, 8, or 0 may be divisible evenly by either odd or even numbers. For example, 12 is evenly divisible by the even numbers 2, 4, and 6, as well as by the odd number 3.

5-11. Test for Divisibility by 2

If a number ends in 0, 2, 4, 6, or 8, it is, of course, evenly divisible by 2.

5-12. Test for Divisibility by 3

If the sum of a number's digits is evenly divisible by 3, the number can be evenly divided by 3.

Example: 573

Test: Digit sum is $5 + 7 + 3 = 15$. Since 15 is evenly divisible by 3, so is the number 573.

Check:

$$3 \overline{)\,573} \\ 191$$

Since changing the order of the digits does not change the digit sum, it follows that 375, 357, 537, 735, and 753 are also evenly divisible by 3.

Instead of getting the sum of all the digits, you can make the test faster and easier by omitting 3's, 6's, and 9's, and combinations that add up to 3, 6, or 9 (2 and 1, 2 and 4, 4 and 5, etc.), and testing the sum of the other digits.

Example: 1,945,369

Test: By omitting the two 9's, the 4 and 5, and the 3 and 6, only the 1 remains. This not only shows that the number is not evenly divisible by 3 but also that there will be a remainder of 1 if the number is divided by 3.

Check:

$$3 \overline{)\,1{,}945{,}369} \\ 648{,}456\tfrac{1}{3} \text{ (1 remainder)}$$

5-13. Test for Divisibility by 4

If a number's last two digits are zeros or are evenly divisible by 4, the number can be evenly divided by 4.

Example: 9,537,724

Test: Since the last two digits (24) are evenly divisible by 4, the number itself is evenly divisible by 4.

You can simplify this test further by dividing the last two digits by 2. If the result is an even number, it shows that the last two digits are evenly divisible by 4 and that, therefore, the entire number is, also.

Check: 4) 9,537,724   2) 9,537,724
             2,384,431   2) 4,768,862
                               2,384,431

5-14. Test for Divisibility by 5

If a number ends in 5 or 0, it is evenly divisible by 5.

5-15. Test for Divisibility by 6

If a number is even and its digit sum is evenly divisible by 3, the number can be evenly divided by 6.

Example: 23,754

Test: The number is even and its digit sum, 21, is evenly divisible by 3. Hence, the number can be evenly divided by 6.

Check: 6) 23,754
              3,959

5-16. Test for Divisibility by 7

There is no simple test for divisibility by 7. The best way is to try dividing the number by 7, especially if the number has fewer than six digits. However, you may find the following method interesting and, at times, useful.

Example (1): 1,224,251

Steps in test:

(*a*) Starting from the right, split the number into groups of three digits: 1 224 251. Numbers of 1,000 or more are generally marked off into groups of three by commas, so you don't have to do this separately.

(b) Add the odd groups (first, third, etc.), add the even groups (second, fourth, etc.; 224 stands alone here, so there is nothing to add), and subtract one from the other: $1 + 251 = 252$; $-224 = 28$.

(c) The original number is evenly divisible by 7 if the remainder in (b) is zero or can be evenly divided by 7. Since the remainder is 28, the number can be evenly divided by 7.

Check: 7) 1,224,251
            174,893

You can use this same test for checking divisibility by 11 and 13, also. Since the remainder (28) cannot be evenly divided by either 11 or 13, neither can the original number.

Check: 11) 1,224,251
              111,295$\frac{6}{11}$

                    94173
        13) 1224251
             117
              54
              52
              22
              13
              95
              91
              41
              39
               2 Remainder

Example (2): Test 220,324 for divisibility by 7, 11, or 13.

Test: $324 - 220 = 104$ which is evenly divisible by 13 but not by 7 or 11. Therefore, the original number can be evenly divided by 13 but not by 7 or 11.

Check: 16948   7) 220,324
        13) 220324   31,474$\frac{6}{7}$
             13
             90
             78
            123
            117
             62   11) 220,324
             52   20,029$\frac{5}{11}$
            104
            104

Example (3): Test 676,522 for divisibility by 7, 11, or 13.

Test: 676 – 522 = 154 which is evenly divisible by both 7 and 11 but not by 13. Therefore, the same applies to the original number.

Check:
7) 676,522
  96,646

    52040
13) 676522
    65
    26
    26
    52
    52
    2 Remainder

11) 676,522
   61,502

A separate test for divisibility by 11 is given in Method 5-20.

5-17. Test for Divisibility by 8

If the last three digits of a number are zeros or are evenly divisible by 8, the number can be evenly divided by 8.

Example: 727,648

Test: The last three digits are evenly divisible by 8 (648 ÷ 8 = 81); therefore, the number can be evenly divided by 8.

Check: 8) 727,648
           90,956

You can simplify this test further by dividing the last three digits by 4. If the result is an even number, it shows that the last three digits are evenly divisible by 8 and that the original number is, also. Or, instead of dividing by 4, you can divide twice by 2 to see if you get an even number.

Of course, if the last *two* digits are not evenly divisible by 4, the number is not evenly divisible by either 4 or 8. And keep in mind that, if the number's last digit is odd, you need go no further since the number cannot be evenly divided by *any* even number.

Here is another quick test. A number is evenly divisible by 8 if (*a*) the hundreds digit is even and the last two digits are evenly divisible by 8, such as in 464; or (*b*) the hundreds digit is odd and the last two digits are evenly divisible by 4 but not by 8, such as in 512.

5-18. Test for Divisibility by 9

If a number's digit sum is evenly divisible by 9, the number can be evenly divided by 9.

Examples: 18: digit sum: 9
36: digit sum: 9
666: digit sum: 18
111,111,111: digit sum: 9

Test: The digit sums of all four numbers are evenly divisible by 9; therefore, the numbers themselves are evenly divisible by 9.

Check: 9) 666   9) 111,111,111
          74   12,345,679

Instead of getting the digit sum, it is faster and easier to cast out 9's. If there is no remainder after casting out 9's, the number can be evenly divided by 9.

Example: 993,645

Test: Casting out 9, 9, 3 + 6 (= 9), and 4 + 5 (= 9), there is no remainder. Therefore, the number is evenly divisible by 9.

Check: 9) 993,645
          110,405

However, if there *is* a remainder after casting out 9's, there will be the same remainder when the number is divided by 9.

Example: 2,081

Test: Casting out 8 and 1 (= 9) leaves a remainder of 2. Therefore, there will also be a remainder of 2 when the number is divided by 9.

Check: 9) 2,081
          $231\frac{2}{9}$

5-19. Test for Divisibility by 10

If a number ends in zero, it is, of course, evenly divisible by 10.

Summary of Tests for Divisibility: 2 to 10

Number is Divisible by	If Number's
2	Last digit is even.
3	Digit sum is evenly divisible by 3.
4	Last two digits are 00 or are evenly divisible by 4.
5	Last digit is 5 or 0.
6	Last digit is even and digit sum is evenly divisible by 3.
7	(See Method 5-16)
8	Last three digits are 000 or are evenly divisible by 8.
9	Digit sum is evenly divisible by 9.
10	Last digit is 0.

With regard to these tests, the following should also be noted:

(a) The tests for divisibility by 2, 5, and 10 involve checking only the last digit.

(b) The test for divisibility by 4 depends on the last two digits.

(c) The test for divisibility by 8 depends on the last three digits.

(d) The tests for divisibility by 3, 6, and 9 are based on digit sums; in the case of 6, the number must also be even.

You can extend these tests to certain larger divisors by applying two of the tests to a number. For example, if the number is evenly divisible by both 3 and 4, it is also evenly divisible by 12 (product of 3 × 4); if evenly divisible by 3 and 5, the number is also evenly divisible by 15.

However, this will not work if the factors that make up the larger divisor can be evenly divided into each other. For instance, you cannot test for divisibility by 16 by testing for 2 and 8 because 2 goes into 8 evenly and any number evenly divisible by 8 must also be evenly divisible by 2. Therefore, this does not provide two distinct tests. Of course, if you divide a number by 8 and get an even number as a result, this does tell you that the number is evenly divisible by 16.

Testing for Divisibility by Larger Divisors

5-20. Test for Divisibility by 11

Starting from the left or right, add the digits in the "odd" places and the digits in the "even" places. If the difference between these sums is 0 or is evenly divisible by 11, the number is evenly divisible by 11.

Example (1): 3,861

Steps in test:
(a) Sum of first and third digits from the right: $1 + 8 = 9$
(b) Sum of second and fourth digits from the right: $6 + 3 = 9$
(c) $9 - 9 = 0$; the number is evenly divisible by 11

Short way: $(1 + 8) - (6 + 3) = 0$

Check: 11) 3,861
                351

Example (2): 937,651

Test: $1 + 6 + 3 = 10$; $5 + 7 + 9 = 21$; $-10 = 11$; the number is evenly divisible by 11.

Check: 11) 937,651
                 85,241

Checking three-digit numbers for divisibility by 11 is especially easy by this method. If the sum of the two outside digits equals the middle digit, or differs by 11 from the middle digit, the number is evenly divisible by 11. (Note that the two outside digits are in "odd" places, the middle digit in an "even" place.)

Example (1): 792 and 297

Test: $7 + 2 = 9$, the middle digit; both 792 and 297 are evenly divisible by 11.

Check: 11) 792   11) 297
                72   27

Example (2): 968 and 869

Test: $9 + 8 = 17$; $-6 = 11$; both 968 and 869 are evenly divisible by 11.

Check: 11) 968   11) 869
                88   79

If a number is made up of a repeating digit, it is evenly divisible by 11 if it has an even number of digits.

Examples: 66, 5,555, and 777,777 are all evenly divisible by 11 but 666, 66,666, and 7,777,777 are not.

Check: 11) 5,555   11) 55,555
                  ‾‾‾‾‾   ‾‾‾‾‾‾
                    505   5,050 $\frac{5}{11}$

Another test for divisibility by 11 is included under Method 5-16.

5-21. Test for Divisibility by 12
If a number is evenly divisible by both 3 and 4, it can be evenly divided by 12.

Example: 3,625,020

Steps in test:
(a) Digit sum is 18; number is evenly divisible by 3 (Method 5-12)
(b) Last two digits (20) are evenly divisible by 4; number is evenly divisible by 4 (Method 5-13)
(c) Since the number is evenly divisible by both 3 and 4, it is evenly divisible by 12

Check: 12) 3,625,020
                  ‾‾‾‾‾‾‾‾
                     302,085

5-22. Test for Divisibility by 15
If a number is evenly divisible by both 3 and 5, it can be evenly divided by 15.

Example: 7,335

Steps in test:
(a) Digit sum is 18; number is evenly divisible by 3
(b) Number ends in 5; is evenly divisible by 5
(c) Since the number meets both tests, it is evenly divisible by 15

Check: 489
                 15) 7335
                     ‾‾‾‾
                     60
                     ‾‾
                     133
                     120
                     ‾‾‾
                     135
                     135
                     ‾‾‾

5-23. Test for Divisibility by 18
If a number is evenly divisible by both 2 and 9, it can be evenly divided by 18.

Example: 2,952

Steps in test:
(a) Number ends in 2; is evenly divisible by 2
(b) Digit sum is 18; number is evenly divisible by 9
(c) Since the number meets both tests, it is evenly divisible by 18

Check:

$$\begin{array}{r} 164 \\ 18\overline{)2952} \\ \underline{18} \\ 115 \\ \underline{108} \\ 72 \\ \underline{72} \end{array}$$
or:
$$\begin{array}{r} 2\overline{)2,952} \\ 9\overline{)1,476} \\ 164 \end{array}$$

5-24. Extending the Tests to Divisibility by Larger Divisors

Tests for divisibility can be extended to larger divisors, as shown by the following examples.

Number is Divisible by	If the Number
20	Ends in 0 and the preceding digit is even.
22	Is even and evenly divisible by 11.
24	Is evenly divisible by both 3 and 8.
25	Ends in 00, 25, 50, or 75.
30	Ends in 0 and is evenly divisible by 3.
33	Is evenly divisible by both 3 and 11.
36	Is evenly divisible by both 4 and 9.

Simplifying Dividends and Divisors

5-25. Simplifying the Dividend by Adding or Subtracting

Example (1): $3,196 \div 4$

Solution by steps:
(a) Simplify 3,196 to an easy zero-number by adding 4: $3,196 + 4 = 3,200$
(b) Divide the simplified dividend: $3,200 \div 4 = 800$; which means that there are 800 4's in 3,200, but since one 4 was added, this is 1 too many
(c) Subtract the 1: $800 - 1 = 799$, answer

Short way: $3,196 \div 4 = \dfrac{3,200 - 4}{4} = 800 - 1 = 799$

Proving the answer: $4\overline{)\,3,196}$
$\phantom{4\overline{)\,}}799$

Example (2): $711 \div 9$

Solution by steps:

(a) Digit sum of 711 shows that the number is evenly divisible by 9. Simplify to a zero-number by adding 9: $711 + 9 = 720$

(b) $720 \div 9 = 80$; since a 9 was added, this is one 9 too many, so 1 must be subtracted.

(c) $80 - 1 = 79$, answer

Short way: $711 \div 9 = \dfrac{720 - 9}{9} = 80 - 1 = 79$

Proving the answer: $9\overline{)\,711}$
$\phantom{9\overline{)\,}}79$

Example (3): $558 \div 9$

Solution by steps:

(a) Digit sum of 558 shows that the number is evenly divisible by 9. Subtracting 18 will give a zero-number that is also divisible by 9: $558 - 18 = 540$

(b) $540 \div 9 = 60$; this represents two 9's fewer than were in the original number; therefore 2 must be added.

(c) $60 + 2 = 62$, answer

Short way: $558 \div 9 = \dfrac{540 + 18}{9} = 60 + 2 = 62$

Proving the answer: $9\overline{)\,558}$
$\phantom{9\overline{)\,}}62$

You can use this method even if there is going to be a remainder.

Example: $3,197 \div 4$

Solution by steps:

(a) Test for divisibility by 4 (Method 5-13) shows that there will be a remainder of 1 ($97 \div 4 = 24\frac{1}{4}$)

(b) Adding 3 to the dividend will make it a zero-number evenly divisible by 4: $3,197 + 3 = 3,200$

(c) $3,200 \div 4 = 800$; this is $\frac{3}{4}$ of a 4 too much because of the 3 that was added; therefore, $\frac{3}{4}$ must be subtracted.

(d) $800 - \frac{3}{4} = 799\frac{1}{4}$, answer

Short way: $3{,}197 \div 4 = \dfrac{3{,}200 - 3}{4} = 800 - \frac{3}{4} = 799\frac{1}{4}$

Proving the answer: $\quad 4)\overline{\,3{,}197\,}$
$\qquad\qquad\qquad\qquad\quad 799\frac{1}{4}$

If the dividend is odd and the divisor even, you may find it helpful to make the dividend even, also.

Example (1): $\; 111 \div 2$

Solution by steps:
(a) $111 - 1 = 110$; the 1 subtracted must be added as a remainder of $\frac{1}{2}$ in the answer.

(b) $110 \div 2 = 55; + \frac{1}{2} = 55\frac{1}{2}$, answer

Short way: $\; 111 \div 2 = \dfrac{110 + 1}{2} = 55\frac{1}{2}$

Example (2): $\; 69 \div 4$

Solution: $\; 69 - 1 = 68; \div 4 = 17; + \frac{1}{4} = 17\frac{1}{4}$

5-26. Simplifying the Dividend by Breaking It into Parts

Example: $168 \div 3$ (test by Method 5-12 shows 168 is evenly divisible by 3)

Solution by steps:
(a) Break the dividend into easier numbers that are also evenly divisible by 3: $168 = 150 + 18$

(b) $\frac{150}{3} + \frac{18}{3} = 50 + 6 = 56$, answer

Breaking the dividend into parts by units, tens, hundreds, etc., is another way to simplify it for easier division.

Example: $4{,}264 \div 8$ (test by Method 5-17 shows number is evenly divisible by 8)

Solution: $4{,}264 = 4{,}000 + 200 + 64; \div 8 = 500 + 25$
$+ 8 = 533$, answer

Proving the answer: $\quad 8)\overline{\,4{,}264\,}$
$\qquad\qquad\qquad\qquad\quad 533$

5-27. Simplifying the Divisor by Factoring It

Factoring the divisor simply means breaking it into factors— into two or more smaller numbers which, when multiplied together, give the divisor as the product, such as breaking 6 into

2 × 3. You may find it easier to divide by such factors than by the larger divisor.

In dividing by factors, it is usually better to start with the smallest one (or smaller one) since this will reduce the dividend for easier division by the larger factors. However, you may sometimes find it easier to start with the larger factor, such as in dividing 81 by 18. After factoring 18 into 2 × 9, it is obviously better to divide first by 9 rather than by 2, thus: $81 \div 9 = 9$; $\div 2 = 4\frac{1}{2}$.

In considering which factor to start with, test the dividend for divisibility and try to divide first by a factor that will not leave a remainder, if you have such a choice.

Example (1): $216 \div 27$

Solution by steps:
(*a*) Factor 27 into 3 × 9 (test by Method 5-18 shows that 216 is evenly divisible by 9 and, therefore, also by 3)

(*b*) $216 \div 3 = 72$; $\div 9 = 8$, answer

Proving the answer:

$$\begin{array}{r} 8 \\ 27\overline{)216} \\ \underline{216} \end{array}$$

also:

$$\begin{array}{r} 3\overline{)216} \\ 9\overline{)\;72} \\ 8 \end{array}$$

Why it works: Dividing by the factors of a number is the same as dividing by the number which is a product of those factors.

Thus, $\dfrac{24}{12} = 2$; $\dfrac{24}{3 \times 4} = \dfrac{8}{4} = 2$, or $\dfrac{6}{3} = 2$;

also, $\dfrac{24}{2 \times 6} = \dfrac{12}{6} = 2$, or $\dfrac{4}{2} = 2$

Example (2): $2,184 \div 24$

Solution by steps:
(*a*) 24 can be split into three pairs of factors: 2 × 12, 3 × 8, and 4 × 6. Use any pair you please. Tests for divisibility show that 2,184 can be evenly divided by all of these factors, and also by 24.

(*b*) $2,184 \div 2 = 1,092$; $\div 12 = 91$, answer
$2,184 \div 3 = 728$; $\div 8 = 91$
$2,184 \div 4 = 546$; $\div 6 = 91$

Proving the answer:

$$
\begin{array}{r}
91 \\
24{\overline{\smash{\big)}\,2184}} \\
\underline{216} \\
24 \\
\underline{24}
\end{array}
$$

Dividing by factors may enable you to simplify a problem from "long" division to easy "short" divisions.

Example: 2,679,568 ÷ 16

Solution by short divisions:
(a) Factor 16 into 2 × 8 or 4 × 4 and divide by either pair of factors.

(b)

$$
\begin{array}{l}
2{\overline{\smash{\big)}\,2,679,568}} \\
8{\overline{\smash{\big)}\,1,339,784}} \\
167,473
\end{array}
\qquad
\begin{array}{l}
4{\overline{\smash{\big)}\,2,679,568}} \\
4{\overline{\smash{\big)}\,669,892}} \\
167,473
\end{array}
$$

Solution by long division:

$$
\begin{array}{r}
167473 \\
16{\overline{\smash{\big)}\,2679568}} \\
\underline{16} \\
107 \\
\underline{96} \\
119 \\
\underline{112} \\
75 \\
\underline{64} \\
116 \\
\underline{112} \\
48 \\
\underline{48}
\end{array}
$$

Other examples:
384 ÷ 24: 384 ÷ 2 = 192; ÷ 2 = 96; ÷ 6 = 16
644 ÷ 16: 644 ÷ 2 = 322; ÷ 2 = 161; ÷ 4 = $40\frac{1}{4}$

If a factor is not evenly divisible into a number, and you do not need the exact answer, you can ignore the remainder after each division. This will give you an answer within 1 less than the exact answer, no matter how many factors you divide by.

Example (1): 913 ÷ 27

Solution by steps:
(a) Factor 27 into 9 × 3 (test for divisibility shows that 913 is not evenly divisible by either factor)

(b) 913 ÷ 9 = $101\frac{4}{9}$; drop the $\frac{4}{9}$

(c) $101 \div 3 = 33\frac{2}{3}$, approximate answer

Exact answer:

$$\begin{array}{r} 33 \\ 27\overline{)\,913} \\ \underline{81} \\ 103 \\ \underline{81} \\ 22 \end{array}$$

answer: $33\frac{22}{27}$

The approximate answer is only $\frac{4}{27}$ less than the exact answer: $33\frac{22}{27} - 33\frac{2}{3}\ (33\frac{18}{27}) = \frac{4}{27}$

Example (2): 22,841 $\div$ 132

Solution by steps:

(a) 132 may be factored into $2 \times 2 \times 3 \times 11$ (tests for divisibility show that 22,841 is not evenly divisible by any of these factors)

(b) In the following divisions, the remainders are ignored and not included in the next division: $22,841 \div 2 = 11,420\frac{1}{2}$; $\div 2 = 5,710$; $\div 3 = 1,903\frac{1}{3}$; $\div 11 = 173$, approximate answer

Exact answer:

$$\begin{array}{r} 173 \\ 132\overline{)\,22841} \\ \underline{132} \\ 964 \\ \underline{924} \\ 401 \\ \underline{396} \\ 5 \end{array}$$

answer: $173\frac{5}{132}$

The approximate answer is only $\frac{5}{132}$ less than the exact answer.

5-28. Simplifying the Dividend and Divisor by Multiplying or Dividing Both by the Same Number

If the divisor ends in 5, doubling it and the dividend can simplify the division by changing the divisor into a number ending in zero.

Example (1): 138 $\div$ 15

Solution: Double both numbers and divide:
$$276 \div 30 = 27.6 \div 3 = 9.2, \text{ answer}$$

Proving the answer:

$$9.2$$
$$15 \overline{)\ 138.0}$$
$$\underline{135}$$
$$30$$
$$\underline{30}$$

Example (2): $560 \div 35$

Solution: $1,120 \div 70 = 112 \div 7 = 16$, answer

Alternate solution: Simplify both numbers first by dividing by 7:

$$\frac{560}{7} \div \frac{35}{7} = 80 \div 5 = 16$$

Proving the answer:

$$16$$
$$35 \overline{)\ 560}$$
$$\underline{35}$$
$$210$$
$$\underline{210}$$

Why it works: The dividend and divisor can be multiplied by the same number, or be divided by the same number, without changing the answer. You can easily see how this works by using small numbers, such as $12 \div 6 = 2$. Halve both numbers: $6 \div 3 = 2$; double both numbers: $24 \div 12 = 2$; triple both numbers: $36 \div 18 = 2$, etc.

5-29. Problems Involving Both Multiplication and Division

If a problem involves both multiplication and division, look it over to see which operation will be easier to do first. Also, whether the multiplier and divisor can be used to cancel or reduce each other.

Example (1): Multiply 484 by 16 and divide by 32

Solution: $16 \div 32 = \frac{1}{2}$; $\times 484 = 242$, answer

also: $\dfrac{484 \times \cancel{16}}{\underset{2}{\cancel{32}}} = 242$

Proving the answer:

$$484$$
$$\times\ 16$$
$$2904$$
$$\underline{484}$$
$$7744 \div 32$$

$$\begin{array}{r} 242 \\ 32\overline{)\,7744} \\ \underline{64} \\ 134 \\ \underline{128} \\ 64 \\ \underline{64} \end{array}$$

Example (2): Divide 336 by 21 and multiply by 45

Solution: $\dfrac{\overset{48}{\cancel{336}} \times \overset{15}{\cancel{45}}}{\underset{3}{\cancel{21}}} = 720$, answer ($48 \times 15 = 24 \times 30 = 720$)

also: $\dfrac{\overset{16}{\underset{\cancel{48}}{\cancel{336}}} \times 45}{\underset{3}{\cancel{21}}} = 720$ ($16 \times 45 = 8 \times 90 = 720$)

Why it works: When multiplication and division are to be done together, it does not matter in what order you perform the operations. You can easily see this by using small numbers, such as in multiplying 3 by 4 and dividing by 6.

$$3 \times 4 = 12; \div 6 = 2$$
$$3 \div 6 = \tfrac{1}{2}; \times 4 = 2$$
$$4 \div 6 = \tfrac{2}{3}; \times 3 = 2$$

Checking Results in Division

5-30. Estimate the Answer Before You Divide

Making an estimate before you divide can help to avoid big mistakes, such as getting an answer in the thousands when it should be in the hundreds, or vice versa. If you get an answer that is very different in magnitude from your estimate, check to see what is wrong.

One way to make an estimate is to round off the dividend and the divisor for easy division. In rounding off, change both numbers in the same direction; that is, either increase both or decrease both. (In rounding off for multiplication, you would increase one number and decrease the other, as mentioned in Method 4-42.)

Example: 45,496 ÷ 517

Estimate: 45,000 ÷ 500 = 450 ÷ 5 = 90, approximate
answer

Exact answer:

$$\begin{array}{r} 88 \\ 517\overline{)45496} \\ \underline{4136} \\ 4136 \\ \underline{4136} \end{array}$$

In some problems, an estimate may be all you need; this can
save you unnecessary figuring. It all depends upon the degree of
accuracy you require for your purpose.

5-31. Check the Subtractions in Long Division As You Go

As you complete each subtraction in long division, add the
remainder back to the number above it as a check. This will help
to catch an error before you go on and will save time in the long
run.

Example:

$$\begin{array}{r} 341 \\ 371\overline{)126511} \\ \underline{1113} \\ 1521 \\ \underline{1484} \\ 371 \\ \underline{371} \end{array}$$

1521 As you go, add 152 back to 1113 to get 1265;
1484 add 37 back to 1484 to get 1521.

After you obtain your answer, use one of the following methods
to check it.

5-32. Checking Division by Multiplying Your Answer
by the Divisor

Multiplying your answer (quotient) by the divisor should
give the dividend as the product.

Example:

$$\begin{array}{r} 34 \text{ Quotient} \\ \text{Divisor } 48\overline{)1632} \text{ Dividend} \\ \underline{144} \\ 192 \\ \underline{192} \end{array}$$

Check:

$$\begin{array}{r} 34 \\ \times\ 48 \\ \hline 272 \\ \underline{136} \\ 1632 \end{array}$$

Multiplying the divisor by the quotient (48 × 34) will also give the dividend as the product but this check is not as good since the same multiplications will be done as in the division and a mistake in multiplying may be repeated.

$$\begin{array}{r} 48 \\ \times\ 34 \\ \hline 192 \\ 144 \\ \hline 1632 \end{array}$$ When the check is made this way, the multiplications, 4 × 48 and 3 × 48, are the same as in the division.

If there is a remainder in the division, add this to the product in checking your answer.

Example:

$$35\overline{)1626}$$ quotient 46
$$\begin{array}{r} \underline{140} \\ 226 \\ \underline{210} \\ 16 \end{array}$$ Remainder

Check:

$$\begin{array}{r} 46 \\ \times\ 35 \\ \hline 230 \\ 138 \\ \hline 1610 \\ +\ 16 \\ \hline 1626 \end{array}$$

5-33. Checking Division by Dividing Your Answer into the Dividend

Dividing your answer (quotient) into the dividend should give the divisor as the new quotient.

Example:

$$38\overline{)2812}$$ quotient 74
$$\begin{array}{r} \underline{266} \\ 152 \\ \underline{152} \end{array}$$

Check:

$$74\overline{)2812}$$ quotient 38
$$\begin{array}{r} \underline{222} \\ 592 \\ \underline{592} \end{array}$$

5-34. Checking Division by Using a Short-Cut

Any short-cut that could be used to simplify the division may be used to check the answer.

Example (1):

$$18\overline{)62208}$$ quotient 3456
$$\begin{array}{r} \underline{54} \\ 82 \\ \underline{72} \\ 100 \\ \underline{90} \\ 108 \\ \underline{108} \end{array}$$

Check:
$$2\overline{)62208}$$
$$9\overline{)31104}$$
$$3456$$

Example (2):

```
        539.448     Check:  67,431 × 8 = 539,448;
125) 67431.000              ÷ 1,000 = 539.448
     625                    (by Method 5-8D)
     493
     375
     1181
     1125
      560
      500
      600
      500
      1000
      1000
```

If you use a short-cut in doing the division the first time, you can check your answer by another short-cut or by conventional division.

5-35. Checking Division by Casting Out 9's

Casting out 9's is probably the fastest and easiest way to check division if the numbers are large. You can cast out 9's by any of the methods described in 1-6, Chapter I.

Small numbers are used in the next example to make it easier to show how the remainders from casting out 9's are applied to checking division.

Example (1):

```
            311
      24) 7468
          72
          26
          24
          28
          24
           4
```

Check by steps:

(a) Cast out 9's from the divisor (24):   6 Remainder

(b) Cast out 9's from the quotient (311):   5 Remainder

(c) Multiply the remainders of (a) and (b):
     $6 \times 5 = 30$; cast out 9's: 3 Remainder

(d) Cast out 9's from the remainder of the
     division (4) and add to the remainder of
     (c): $4 + 3 = 7$:   7 Remainder

(*e*) Cast out 9's from the dividend (7468):   7 Remainder

The remainder (7) is the same in (*d*) and (*e*); the answer may therefore be assumed to be correct.

Example (2):

$$
\begin{array}{r}
1897 \\
4364)\overline{8279461} \\
\underline{4364} \\
39154 \\
\underline{34912} \\
42426 \\
\underline{39276} \\
31501 \\
\underline{30548} \\
953
\end{array}
$$

Check by casting out 9's

(*a*) Remainder of divisor: 8

(*b*) Remainder of quotient: 7

(*c*) $8 \times 7 = 56$; remainder: 2

(*d*) Remainder of division's remainder (953): 8; add to (*c*): $8 + 2 = 10$; remainder: 1

(*e*) Remainder of dividend: 1

Since the remainder (1) is the same in (*d*) and (*e*), the answer may be considered correct.

Keep in mind that a check by casting out 9's will not catch an error caused by digits being transposed or a decimal point misplaced since such errors will not affect the remainders. However, if the check does *not* prove your answer, there is definitely something wrong in the division, in the check, or in both.

VI. FRACTIONS, MIXED NUMBERS, AND PERCENTAGES

6-1. Definitions and Basic Principles
Common fractions:

A common fraction is written with a whole number above a line and a whole number below the line, thus: $\dfrac{3}{4}\ \begin{matrix}\text{Numerator}\\\text{Denominator.}\end{matrix}$ Common fractions are usually referred to simply as "fractions."

Fractions are a form of division. The line of a fraction means to divide the upper number (numerator) by the lower number (denominator); it has the same meaning as a division sign: $\frac{3}{4}$ or $3/4 = 3 \div 4 = 4)\overline{3}$. When "of" is used with a fraction, it means to multiply: $\frac{3}{4}$ of $8 = \frac{3}{4} \times 8 = 6$.

Multiplying the numerator and denominator by the same number, or dividing both by the same number, does not change the value of a fraction, as shown here:

$$\frac{2 \times 4}{5 \times 4} = \frac{8}{20} \div \frac{2}{2} = \frac{4}{10} \div \frac{2}{2} = \frac{2}{5}$$

Multiplying by $\frac{4}{4}$ is the same as multiplying by 1; dividing by $\frac{2}{2}$ is the same as dividing by 1. Multiplying or dividing by 1 leaves the value of the fraction unchanged.

Decimal fractions:

A decimal fraction is a common fraction that has 10, 100, 1,000, etc., as the denominator.

Examples: $\frac{1}{10}, \frac{1}{100}, \frac{1}{1000}$

Such fractions may be expressed in decimal form by using a decimal point. Thus, $\frac{1}{10} = .1$, $\frac{1}{100} = .01$, $\frac{1}{1000} = .001$, etc. Fractions written this way are usually referred to as "decimals."

As you know, the value of a digit is decreased by one-tenth

for each place to the right of the decimal point and is increased ten times for each place to the left of the decimal point.

In general, you can use the same short-cuts with decimals as with whole numbers, but be careful to keep decimal points in their proper places. When multiplying with decimals, there must be as many decimal places in the product as the total of decimal places in the factors.

Mixed numbers:

A mixed number is a whole number with a fraction, such as $1\frac{1}{2}$, $2\frac{1}{4}$, $3\frac{3}{4}$. Such numbers may be reduced to fractional form, thus: $1\frac{1}{2} = \frac{3}{2}$; $2\frac{1}{4} = \frac{9}{4}$; $3\frac{3}{4} = \frac{15}{4}$.

Adding and Subtracting Fractions

6-2. Finding a Common Denominator

Before fractions can be added or subtracted, they must have the same denominator—a common denominator that is evenly divisible by the denominator of each fraction.

Example of addition: $\frac{1}{4} + \frac{2}{3}$

Solution by steps:

(a) Since "fourths" and "thirds" cannot be added directly, obtain a common denominator by multiplying the denominators: $4 \times 3 = 12$

(b) Change the fractions to the common denominator: $\frac{1}{4} = \frac{3}{12}$; $\frac{2}{3} = \frac{8}{12}$. Now you can add "twelfths".

(c) Add the numerators and place their sum over the common denominator: $3 + 8 = 11$ twelfths $= \frac{11}{12}$, answer

Other examples:

$\frac{1}{3} + \frac{1}{5} + \frac{3}{8}$: $3 \times 5 \times 8 = 120$, common denominator;

$\frac{40}{120} + \frac{24}{120} + \frac{45}{120} = \frac{109}{120}$, answer

$\frac{1}{9} + \frac{3}{4} + \frac{1}{7}$: $9 \times 4 \times 7 = 252$, common denominator;

$\frac{28}{252} + \frac{189}{252} + \frac{36}{252} = \frac{253}{252}$, or $1\frac{1}{252}$, answer

Example of subtraction: $\frac{2}{5} - \frac{1}{4}$

Solution by steps:

(a) Common denominator: $5 \times 4 = 20$

(b) Change fractions to common denominator:

$\frac{2}{5} = \frac{8}{20}$; $\frac{1}{4} = \frac{5}{20}$

(c) Subtract: $\frac{8}{20} - \frac{5}{20} = \frac{3}{20}$, answer

also:
$$\frac{8-5}{20} = \frac{3}{20}$$

Other examples: $\frac{1}{6} - \frac{1}{9} = \frac{9}{54} - \frac{6}{54} = \frac{3}{54} = \frac{1}{18}$, answer

$\frac{3}{5} - \frac{2}{7} = \frac{21}{35} - \frac{10}{35} = \frac{11}{35}$, answer

6-3. Converting Fractions to Decimals for Easier Adding or Subtracting

If fractions have denominators of 2, 4, 5, 8, or 10, you can change them easily into decimals to simplify adding or subtracting. This avoids the need to find a common denominator.

Example: $\frac{1}{5} + \frac{3}{4}$

Solution: $\frac{1}{5} = .2$; $\frac{3}{4} = .75$; $+.2 = .95$, or $\frac{95}{100}$,

or $\frac{19}{20}$, answer

also: $\frac{1}{5} + \frac{3}{4} = \frac{4}{20} + \frac{15}{20} = \frac{19}{20}$

Other examples:

$\frac{1}{8} + \frac{2}{5} = .125 + .4 = .525 \ (\frac{525}{1000}$ or $\frac{21}{40})$

$\frac{1}{2} - \frac{1}{5} = .5 - .2 = .3 \ (\frac{3}{10})$

$\frac{3}{10} - \frac{1}{4} = .3 - .25 = .05 \ (\frac{5}{100}$ or $\frac{1}{20})$

Fractions with denominators of 3, 6, 7, or 9 cannot be converted into decimals that come out "even." For example, $\frac{1}{3} = .3333 \ldots$ $\frac{1}{3}$; $\frac{1}{6} = .1666 \ldots \frac{2}{3}$, etc. It is therefore easier to add and subtract such fractions in their fractional form.

6-4. Fastest Way to Add Any Pair of Fractions

Example: $\frac{1}{4} + \frac{2}{3}$

Solution by steps:

(a) Multiply the denominators: $4 \times 3 = 12$, new denominator

(b) Cross multiply the numerators and denominators, and add: $1 \times 3 = 3$; $4 \times 2 = 8$; $+3 = 11$, new numerator

(c) Answer: $\frac{11}{12}$

Short way: $\frac{1}{4} + \frac{2}{3} = \frac{3+8}{4 \times 3} = \frac{11}{12}$

Other examples: $\frac{1}{6} + \frac{2}{5} = \frac{17}{30}$

$\frac{4}{9} + \frac{3}{8} = \frac{59}{72}$

6-5. Fastest Way to Subtract Any Pair of Fractions

Example: $\frac{2}{5} - \frac{1}{4}$

Solution by steps:

(a) Multiply the denominators (same step as in adding fractions): $5 \times 4 = 20$, new denominator

(b) Cross multiply the numerators and denominators (as in adding), and subtract: $5 \times 1 = 5$; $4 \times 2 = 8$; $- 5 = 3$, new numerator

(c) Answer: $\frac{3}{20}$

Short way: $\frac{2}{5} - \frac{1}{4} = \frac{8 - 5}{5 \times 4} = \frac{3}{20}$

Other examples: $\frac{5}{7} - \frac{3}{11} = \frac{34}{77}$

$\frac{11}{12} - \frac{5}{8} = \frac{28}{96} = \frac{7}{24}$

6-6. To Add Two Fractions with Numerators of 1

The sum of the fractions $= \dfrac{\text{sum of the denominators}}{\text{product of the denominators}}$

Example: $\frac{1}{5} + \frac{1}{6}$

Solution: $\frac{5 + 6}{5 \times 6} = \frac{11}{30}$, answer

Proving the answer: $\frac{1}{5} = \frac{6}{30}$; $\frac{1}{6} = \frac{5}{30}$; $\frac{6}{30} + \frac{5}{30} = \frac{11}{30}$

Other examples: $\frac{1}{8} + \frac{1}{7} = \frac{15}{56}$

$\frac{1}{3} + \frac{1}{11} = \frac{14}{33}$

6-7. To Subtract Two Fractions with Numerators of 1

The difference of the fractions $= \dfrac{\text{difference of the denominators}}{\text{product of the denominators}}$

Example: $\frac{1}{5} - \frac{1}{6}$

Solution: $\frac{6 - 5}{6 \times 5} = \frac{1}{30}$, answer

Proving the answer: $\frac{1}{5} = \frac{6}{30}$; $\frac{1}{6} = \frac{5}{30}$; $\frac{6}{30} - \frac{5}{30} = \frac{1}{30}$

Other examples: $\frac{1}{7} - \frac{1}{9} = \frac{2}{63}$

$\frac{1}{3} - \frac{1}{11} = \frac{8}{33}$

Multiplying and Dividing by Fractions

6-8. To Multiply by a Fraction

Multiplying is probably the easiest of the operations with fractions.

A. To multiply a whole number by a fraction, divide the number by the denominator and multiply by the numerator; or multiply first by the numerator and then divide by the denominator, whichever way is easier.

Examples: $15 \times \frac{2}{3}$: $15 \div 3 = 5$; $\times 2 = 10$

or: $15 \times 2 = 30$; $\div 3 = 10$

$16 \times \frac{3}{4}$: $16 \div 4 = 4$; $\times 3 = 12$

$23 \times \frac{3}{10}$: $23 \times 3 = 69$; $\div 10 = 6.9$

B. To multiply fractions together, multiply numerators by numerators and denominators by denominators but first simplify by reducing and cancelling, where possible.

Example: $\frac{3}{5} \times \frac{1}{6} \times \frac{10}{11}$

Solution: $\frac{3}{5} \times \frac{1}{6} \times \frac{10}{11} = \frac{1}{11}$, answer

Proving the answer: $\frac{3 \times 1 \times 10}{5 \times 6 \times 11} = \frac{30}{330} = \frac{1}{11}$

Other examples:

$\frac{3}{7} \times \frac{8}{5} \times \frac{4}{13} = \frac{3 \times 8 \times 4}{7 \times 5 \times 13} = \frac{96}{455}$, answer

$\frac{5}{18} \times \frac{9}{10} \times \frac{3}{4} = \frac{5 \times 9 \times 3}{18 \times 10 \times 4} = \frac{3}{16}$, answer

To save time in simplifying a fraction, apply the tests for divisibility (Methods 5-10 to 5-24) to see if the numerator and denominator can be divided evenly by the same number (a common factor). Keep dividing by common factors until the fraction has been reduced to its simplest form.

6-9. To Divide by a Fraction

Invert the fraction and proceed as in multiplication.

Example: $8 \div \frac{1}{2}$ (when $\frac{1}{2}$ is inverted it becomes $\frac{2}{1}$ or simply 2)

Solution: $8 \div \frac{1}{2} = 8 \times 2 = 16$, answer; which means that $\frac{1}{2}$ goes into 8 16 times, or that there are 16 $\frac{1}{2}$'s in 8

Other examples: $9 \div \frac{3}{4} = \overset{3}{\cancel{9}} \times \dfrac{4}{\cancel{3}} = 12$

$$16 \div \frac{8}{9} = \overset{2}{\cancel{16}} \times \dfrac{9}{\cancel{8}} = 18$$

If fractions have a common denominator, you can ignore the denominators and divide the numerators like whole numbers.

Examples: $\frac{4}{7} \div \frac{2}{7} = 4 \div 2 = 2$, answer; or $\dfrac{\overset{2}{\cancel{4}}}{\cancel{7}} \times \dfrac{\cancel{7}}{\cancel{2}} = 2$

$\frac{2}{7} \div \frac{4}{7} = 2 \div 4 = \frac{1}{2}$, answer; or $\dfrac{\cancel{2}}{\cancel{7}} \times \dfrac{\cancel{7}}{\underset{2}{\cancel{4}}} = \frac{1}{2}$

$\frac{5}{9} \div \frac{2}{9} = 5 \div 2 = 2\frac{1}{2}$, answer

6-10. To Multiply a Number by $\frac{3}{4}$

Method A: Take $\frac{1}{4}$ of the number and subtract it from the number.

Example: $1,848 \times \frac{3}{4}$

Solution: $1,848 \times \frac{1}{4} = 462$; $1,848 - 462 = 1,386$, answer

Proving the answer:

$$\begin{array}{r} 1,848 \\ \times\,3 \\ \hline 4)\,\overline{5,544} \\ \hline 1,386 \end{array}$$

Why it works: Subtracting $\frac{1}{4}$ of a number from the number leaves $\frac{3}{4}$ of the number. This is the same as multiplying the number by $\frac{3}{4}$.

Method B: Halve the number and, to this, add $\frac{1}{2}$ of the half.

Example: $128 \times \frac{3}{4}$

Solution: $128 \times \frac{1}{2} = 64$; $\times \frac{1}{2} = 32$; $+ 64 = 96$, answer

Proving the answer:

$$\begin{array}{r} 128 \\ \times\,3 \\ \hline 4)\,\overline{384} \\ \hline 96 \end{array}$$

Why it works: $\frac{1}{2} + (\frac{1}{2} \text{ of } \frac{1}{2}) = \frac{1}{2} + \frac{1}{4} = \frac{3}{4}$

Multiplying and Dividing with Mixed Numbers

6-11. Simplifying a Mixed Number Before Multiplying

Example: $19\frac{7}{8} \times 16$

Solution by steps:

(a) Simplify $19\frac{7}{8}$ to a whole number ending in zero by adding $\frac{1}{8}$: $= 20$

(b) Multiply: $20 \times 16 = 320$; this is $\frac{1}{8}$ of 16 too much

(c) Subtract $\frac{1}{8}$ of 16: $320 - 2 = 318$, answer

Proving the answer:

$$
\begin{array}{r}
19\frac{7}{8} \\
\times 16 \\
\hline
14 \ (16 \times \tfrac{7}{8} = 14) \\
114 \\
19 \\
\hline
318
\end{array}
$$

Other examples:

$15\frac{3}{4} \times 25$: $16 \times 25 = 400$; $-(\frac{1}{4} \times 25) = 393\frac{3}{4}$

$18\frac{2}{3} \times 39$: $20 \times 39 = 780$; $-(1\frac{1}{3} \times 39) = 728$

You can also simplify a mixed number by splitting it into parts —into the whole number and the fraction—and multiplying these separately.

Example: $19\frac{7}{8} \times 16$

Solution by steps:

(a) $19 \times 16 = 304$ (by Method 4-32 or 4-39)

(b) $\frac{7}{8} \times 16 = 14$

(c) Add (a) and (b): $304 + 14 = 318$, answer

Short way: $19 \times 16 = 304$; $+ (\frac{7}{8} \times 16) = 318$

6-12. To Square a Mixed Number Ending in $\frac{1}{2}$

Example: $4\frac{1}{2} \times 4\frac{1}{2}$ (same as 4.5×4.5 in Method 4-13)

Solution by steps:

(a) Multiply the whole number by the number $+ 1$: $4 \times 5 = 20$

(b) Place $\frac{1}{4}$ after the product: $20\frac{1}{4}$, answer

Short way: $4 \times 5 + \frac{1}{4} = 20\frac{1}{4}$

Proving the answer: $4\frac{1}{2} \times 4\frac{1}{2} = \frac{9}{2} \times \frac{9}{2} = \frac{81}{4} = 20\frac{1}{4}$

Other examples: $6\frac{1}{2} \times 6\frac{1}{2} = 42\frac{1}{4}$

$7\frac{1}{2} \times 7\frac{1}{2} = 56\frac{1}{4}$

$9\frac{1}{2} \times 9\frac{1}{2} = 90\frac{1}{4}$

6-13. To Multiply a Number by a Mixed Number Ending in $\frac{1}{2}$: Halve-and-Double Method

Halve the number to be multiplied, double the mixed number, and multiply. This gets rid of the fraction for easier multiplication.

Example: $48 \times 3\frac{1}{2}$

Solution: $\frac{1}{2} \times 48 = 24$; $2 \times 3\frac{1}{2} = 7$; $\times 24 = 168$, answer

Proving the answer:

$$48 \times 3\frac{1}{2} = \overset{24}{\cancel{48}} \times \frac{7}{\cancel{2}} = 168$$

Why it works: In multiplication, halving one number and doubling the other does not change the product of the two numbers. (Method 4-25 for multiplying by 15, 1.5, 35, 3.5, etc., works the same way.)

Other examples: $64 \times 3\frac{1}{2} = 32 \times 7 = 224$

$22 \times 4\frac{1}{2} = 11 \times 9 = 99$

$4.8 \times 3\frac{1}{2} = 2.4 \times 7 = 16.8$

6-14. To Divide a Number by a Mixed Number Ending in $\frac{1}{2}$

Double both numbers and divide. This gets rid of the fraction for easier division.

Example: $42 \div 3\frac{1}{2}$

Solution: $2 \times 42 = 84$; $2 \times 3\frac{1}{2} = 7$; $84 \div 7 = 12$, answer

Proving the answer: $42 \div 3\frac{1}{2} = 42 \div \frac{7}{2} = \overset{6}{\cancel{42}} \times \frac{2}{\cancel{7}} = 12$

Why it works: Multiplying the dividend and the divisor by the same number does not change the answer in division. (Method 5-28 for dividing by whole numbers ending in 5 works the same way.)

Other examples: $37 \div 4\frac{1}{2} = 74 \div 9 = 8\frac{2}{9}$

$51 \div 5\frac{1}{2} = 102 \div 11 = 9\frac{3}{11}$

6-15. To Multiply Two Mixed Numbers When They Both End in $\frac{1}{2}$ and the Sum of Their Whole Numbers Is Even

Example: $4\frac{1}{2} \times 6\frac{1}{2}$ (same as 4.5×6.5 in Method 4-29)

Solution by steps:

(a) Multiply the whole numbers: $4 \times 6 = 24$

(b) Obtain $\frac{1}{2}$ the sum of the whole numbers: $\dfrac{4+6}{2} = 5$

(c) Add (a) and (b) and attach $\frac{1}{4}$: $24 + 5 = 29$; $29\frac{1}{4}$, answer

Short way: $(4 \times 6) + \frac{1}{2}(4 + 6) = 29$; answer $29\frac{1}{4}$

Proving the answer: $4\frac{1}{2} \times 6\frac{1}{2} = \frac{9}{2} \times \frac{13}{2} = \frac{117}{4} = 29\frac{1}{4}$

Other examples: $8\frac{1}{2} \times 4\frac{1}{2} = 38\frac{1}{4}$

$7\frac{1}{2} \times 3\frac{1}{2} = 26\frac{1}{4}$

$12\frac{1}{2} \times 8\frac{1}{2} = 106\frac{1}{4}$

6-16. To Multiply Two Mixed Numbers When They Both End in $\frac{1}{2}$ and the Sum of Their Whole Numbers Is Odd

Example: $7\frac{1}{2} \times 2\frac{1}{2}$ (same as 7.5 × 2.5 in Method 4-30)

Solution by steps: same as Method 6-15 but there is a fraction to be dropped and the answer ends in $\frac{3}{4}$ instead of $\frac{1}{4}$.

(a) Multiply the whole numbers: $7 \times 2 = 14$

(b) Obtain $\frac{1}{2}$ the sum of the whole numbers: $\dfrac{7 + 2}{2} = 4\frac{1}{2}$;

drop the $\frac{1}{2}$, $= 4$

(c) Add (a) and (b) and attach $\frac{3}{4}$: $14 + 4 = 18$; $18\frac{3}{4}$, answer

Short way: $(7 \times 2) + \frac{1}{2}(7 + 2) - \frac{1}{2} = 18$; answer $18\frac{3}{4}$

Other examples: $5\frac{1}{2} \times 4\frac{1}{2} = 24\frac{3}{4}$

$6\frac{1}{2} \times 9\frac{1}{2} = 61\frac{3}{4}$

$11\frac{1}{2} \times 12\frac{1}{2} = 143\frac{3}{4}$

6-17. To Multiply Two Mixed Numbers Having the Same Whole Numbers, and Fractions That Add to 1

Example: $6\frac{3}{4} \times 6\frac{1}{4}$

Solution by steps:

(a) Multiply the whole number by the number + 1: $6 \times 7 = 42$

(b) Find the product of the fractions: $\frac{3}{4} \times \frac{1}{4} = \frac{3}{16}$

(c) Add (a) and (b): $42 + \frac{3}{16} = 42\frac{3}{16}$, answer

Short way: $(6 \times 7) + (\frac{3}{4} \times \frac{1}{4}) = 42\frac{3}{16}$

Proving the answer: $6\frac{3}{4} \times 6\frac{1}{4} = \frac{27}{4} \times \frac{25}{4} = \frac{675}{16} = 42\frac{3}{16}$

Other examples: $5\frac{1}{8} \times 5\frac{7}{8} = 30\frac{7}{64}$

$7\frac{1}{9} \times 7\frac{8}{9} = 56\frac{8}{81}$

$16\frac{1}{4} \times 16\frac{3}{4} = 272\frac{3}{16}$

6-18. To Multiply or Divide by $2\frac{1}{2}$

A. To multiply a number by $2\frac{1}{2}$, divide the number by 4 and multiply by 10 (same as Method 4-8B for multiplying by 2.5).

Example: $144 \times 2\frac{1}{2}$

Solution: $144 \div 4 = 36$; $\times 10 = 360$, answer

Proving the answer: $144 \times 2\frac{1}{2} = \overset{72}{\cancel{144}} \times \frac{5}{\cancel{2}} = 360$

Why it works: Dividing by 4 and multiplying by 10 is the same as multiplying by $\frac{10}{4}$ which equals $2\frac{1}{2}$.

B. To divide a number by $2\frac{1}{2}$, multiply the number by 4 and divide by 10 (same as Method 5-7B for dividing by 2.5).

Example: $72 \div 2\frac{1}{2}$

Solution: $72 \times 4 = 288$; $\div 10 = 28.8$, answer

Proving the answer: $72 \div 2\frac{1}{2} = 72 \div \frac{5}{2} = 72 \times \frac{2}{5}$
$$= \frac{144}{5} = 28.8$$

Why it works: Multiplying by 4 and dividing by 10 is the same as multiplying by $\frac{4}{10}$, which is equivalent to dividing by $\frac{10}{4}$ which equals $2\frac{1}{2}$.

Other examples: $635 \div 2\frac{1}{2} = 254$
$24 \div 2\frac{1}{2} = 9.6$

6-19. To Multiply or Divide by $12\frac{1}{2}$

A. To multiply a number by $12\frac{1}{2}$, divide the number by 8 and multiply by 100 (same as Method 4-9C for multiplying by 12.5).

Example: $168 \times 12\frac{1}{2}$

Solution: $168 \div 8 = 21$; $\times 100 = 2,100$, answer

Proving the answer: $168 \times 12\frac{1}{2} = \overset{84}{\cancel{168}} \times \frac{25}{\cancel{2}} = 2,100$

Why it works: Dividing by 8 and multiplying by 100 is the same as multiplying by $\frac{100}{8}$ which equals $12\frac{1}{2}$.

Other examples:
   $38 \times 12\frac{1}{2}$: $38 \div 8 = 4\frac{3}{4} = 4.75$; $\times 100 = 475$
   $81 \times 12\frac{1}{2}$: $81 \div 8 = 10\frac{1}{8} = 10.125$; $\times 100 = 1,012.5$

B. To divide a number by $12\frac{1}{2}$, multiply the number by 8 and divide by 100 (same as Method 5-8C for dividing by 12.5).

Example: $52 \div 12\frac{1}{2}$

Solution: $52 \times 8 = 416; \div 100 = 4.16$, answer

Proving the answer:
$$52 \div 12\tfrac{1}{2} = 52 \div \tfrac{25}{2} = 52 \times \tfrac{2}{25} = \tfrac{104}{25} = 4\tfrac{4}{25} = 4.16$$

Why it works: Multiplying by 8 and dividing by 100 is the same as multiplying by $\tfrac{8}{100}$, which is equivalent to dividing by $\tfrac{100}{8}$ which equals $12\tfrac{1}{2}$.

Other examples: $83 \div 12\tfrac{1}{2}$: $83 \times 8 = 664; \div 100 = 6.64$
$315 \div 12\tfrac{1}{2}$: $315 \times 8 = 2,520; \div 100 = 25.2$

6-20. To Multiply or Divide by $16\tfrac{2}{3}$

A. To multiply a number by $16\tfrac{2}{3}$, divide the number by 6 and multiply by 100.

Example: $72 \times 16\tfrac{2}{3}$

Solution: $72 \div 6 = 12; \times 100 = 1,200$, answer

Proving the answer: $72 \times 16\tfrac{2}{3} = \overset{24}{\cancel{72}} \times \dfrac{\overset{50}{}}{\underset{3}{\cancel{3}}} = 1,200$

Why it works: Dividing by 6 and multiplying by 100 is the same as multiplying by $\tfrac{100}{6}$ which equals $16\tfrac{2}{3}$.

Other examples:
$39 \times 16\tfrac{2}{3}$: $39 \div 6 = 6\tfrac{1}{2} = 6.5; \times 100 = 650$
$44 \times 16\tfrac{2}{3}$: $44 \div 6 = 7\tfrac{1}{3}; \times 100 = 733\tfrac{1}{3}$

B. To divide a number by $16\tfrac{2}{3}$, multiply the number by 6 and divide by 100.

Example: $91 \div 16\tfrac{2}{3}$

Solution: $91 \times 6 = 546; \div 100 = 5.46$, answer

Proving the answer:
$$91 \div 16\tfrac{2}{3} = 91 \div \tfrac{50}{3} = 91 \times \tfrac{3}{50} = \tfrac{273}{50} = 5.46$$

Why it works: Multiplying by 6 and dividing by 100 is the same as multiplying by $\tfrac{6}{100}$, which is equivalent to dividing by $\tfrac{100}{6}$ which equals $16\tfrac{2}{3}$.

Other examples:
$39 \div 16\tfrac{2}{3}$: $39 \times 6 = 234; \div 100 = 2.34$
$141 \div 16\tfrac{2}{3}$: $141 \times 6 = 846; \div 100 = 8.46$

6-21. To Multiply or Divide by $33\tfrac{1}{3}$

A. To multiply a number by $33\tfrac{1}{3}$, divide the number by 3 and multiply by 100.

Example: $93 \times 33\frac{1}{3}$

Solution: $93 \div 3 = 31$; $\times 100 = 3{,}100$, answer

Proving the answer: $93 \times 33\frac{1}{3} = \overset{31}{\cancel{93}} \times \frac{100}{\cancel{3}} = 3{,}100$

Why it works: Dividing by 3 and multiplying by 100 is the same as multiplying by $\frac{100}{3}$ which equals $33\frac{1}{3}$.

Other examples: $28 \times 33\frac{1}{3}$: $28 \div 3 = 9\frac{1}{3}$; $\times 100 = 933\frac{1}{3}$

$47 \times 33\frac{1}{3}$: $47 \div 3 = 15\frac{2}{3}$; $\times 100 = 1{,}566\frac{2}{3}$

B. To divide a number by $33\frac{1}{3}$, multiply the number by 3 and divide by 100.

Example: $53 \div 33\frac{1}{3}$

Solution: $53 \times 3 = 159$; $\div 100 = 1.59$, answer

Proving the answer:
$53 \div 33\frac{1}{3} = 53 \div \frac{100}{3} = 53 \times \frac{3}{100} = \frac{159}{100} = 1.59$

Why it works: Multiplying by 3 and dividing by 100 is the same as multiplying by $\frac{3}{100}$, which is equivalent to dividing by $\frac{100}{3}$ which equals $33\frac{1}{3}$.

Other examples: $67 \div 33\frac{1}{3}$: $67 \times 3 = 201$; $\div 100 = 2.01$

$156 \div 33\frac{1}{3}$: $156 \times 3 = 468$; $\div 100 = 4.68$

6-22. To Multiply or Divide by $66\frac{2}{3}$

A. To multiply a number by $66\frac{2}{3}$, divide the number by 3 and multiply by 200 (or by 2, then by 100).

Example: $48 \times 66\frac{2}{3}$

Solution: $48 \div 3 = 16$; $\times 200 = 3{,}200$, answer

Proving the answer: $48 \times 66\frac{2}{3} = \overset{16}{\cancel{48}} \times \frac{200}{\cancel{3}} = 3{,}200$

Why it works: Dividing by 3 and multiplying by 200 is the same as multiplying by $\frac{200}{3}$ which equals $66\frac{2}{3}$.

Other examples:

$37 \times 66\frac{2}{3}$: $37 \div 3 = 12\frac{1}{3}$; $\times 2 = 24\frac{2}{3}$; $\times 100 = 2{,}466\frac{2}{3}$

$18.63 \times 66\frac{2}{3}$: $18.63 \div 3 = 6.21$; $\times 2 = 12.42$; $\times 100$
$= 1{,}242$

Alternate method of multiplying by $66\frac{2}{3}$:
Multiply the number by 100 and subtract $\frac{1}{3}$ of the product.
Example: $48 \times 66\frac{2}{3}$

Solution: 48 × 100 = 4,800; − 1,600 = 3,200, answer

Why it works: $66\frac{2}{3}$ is $\frac{1}{3}$ less than 100 so that multiplying by 100 and subtracting $\frac{1}{3}$ of the product is the same as multiplying by $66\frac{2}{3}$.

B. To divide a number by $66\frac{2}{3}$, multiply the number by 3 and divide by 200 (or by 2, then by 100).

Example: $46 \div 66\frac{2}{3}$

Solution: 46 × 3 = 138; ÷ 2 = 69; ÷ 100 = .69, answer

Proving the answer:

$46 \div 66\frac{2}{3} = 46 \div \frac{200}{3} = 46 \times \frac{3}{200} = \frac{138}{200} = \frac{69}{100} = .69$

Why it works: Multiplying by 3 and dividing by 200 is the same as multiplying by $\frac{3}{200}$, which is equivalent to dividing by $\frac{200}{3}$ which equals $66\frac{2}{3}$.

Other examples:

$154 \div 66\frac{2}{3}$: 154 × 3 = 462; ÷ 2 = 231; ÷ 100 = 2.31

$47 \div 66\frac{2}{3}$: 47 × 3 = 141; ÷ 2 = 70.5; ÷ 100 = .705

Alternate method of dividing by $66\frac{2}{3}$:

Divide the number by 100 and increase the result by half.

Example: $284 \div 66\frac{2}{3}$

Solution:

284 ÷ 100 = 2.84; × $\frac{1}{2}$ = 1.42; + 2.84 = 4.26, answer

Proving the answer:

$$284 \div 66\frac{2}{3} = 284 \div \frac{200}{3} = \overset{142}{\cancel{284}} \times \frac{3}{\underset{100}{\cancel{200}}} = \frac{426}{100} = 4.26$$

Why it works: 100 is $\frac{1}{2}$ greater than $66\frac{2}{3}$ ($66\frac{2}{3} + \frac{1}{2}$ of $66\frac{2}{3}$ = $66\frac{2}{3} + 33\frac{1}{3}$ = 100) so that dividing a number by 100 and adding $\frac{1}{2}$ of the result gives the same answer as dividing the number by $66\frac{2}{3}$.

Table of Multiplying by Certain Mixed Numbers

N stands for the Number to be multiplied

To Multiply by	Method
$2\frac{1}{2}$	N/4 × 10
$12\frac{1}{2}$	N/8 × 100
$16\frac{2}{3}$	N/6 × 100
$33\frac{1}{3}$	N/3 × 100
$37\frac{1}{2}$	N/8 × 3 × 100
$62\frac{1}{2}$	N/8 × 5 × 100
$66\frac{2}{3}$	N/3 × 2 × 100
$66\frac{2}{3}$	100N − 100N/3
$87\frac{1}{2}$	100N − 100N/8

Table of Dividing by Certain Mixed Numbers

N stands for the Number to be divided

To Divide by	Method
$2\frac{1}{2}$	4N/10
$12\frac{1}{2}$	8N/100
$16\frac{2}{3}$	6N/100
$33\frac{1}{3}$	3N/100
$37\frac{1}{2}$	8N/3 ÷ 100
$62\frac{1}{2}$	8N/5 ÷ 100
$66\frac{2}{3}$	3N/2 ÷ 100
$66\frac{2}{3}$	N/100 + ($\frac{1}{2}$ × N/100)
$87\frac{1}{2}$	(N + N/7) ÷ 100

Working with Percentages

"Per cent" simply means "per hundred" or "per 100". Thus, 6% means 6 per 100, or 6 in every 100, or 6 out of every 100. This may also be written as a decimal, .06, or as a common fraction, $\frac{6}{100}$.

Accordingly, 1% of a number is $\frac{1}{100}$ of the number, 10% of a number is $\frac{10}{100}$ or $\frac{1}{10}$ of the number, 50% of a number is $\frac{50}{100}$ or $\frac{1}{2}$ of the number, etc.

To answer the question as to what per cent one number is of another, simply set up the numbers as a fraction and change this to per cent, as follows:

1. Make the number that follows the words, "what per cent is," the numerator of the fraction.

2. Make the other number the denominator of the fraction.

3. Reduce the fraction, if possible, and then change it to per cent by multiplying by 100.

Example: What per cent is 16 of 64?

Solution by steps:

(a) Make 16 the numerator and 64 the denominator: $\frac{16}{64}$; this can be reduced to $\frac{1}{4}$

(b) Multiply the fraction by 100 to change it to per cent: $\frac{1}{4}$ × 100 = 25%; which means, of course, that 16 is 25% of 64

Short way: $\frac{16}{64} = \frac{1}{4}$; × 100 = 25%

Other examples:

What per cent is 64 of 16? $\frac{64}{16} = 4$; × 100 = 400%

What per cent is 5 of 10? $\frac{5}{10} = \frac{1}{2}$; × 100 = 50%

What per cent is 4 of 36? $\frac{4}{36} = \frac{1}{9}$; × 100 = $11\frac{1}{9}$%

6-23. To Find $16\frac{2}{3}\%$ of a Number

Divide the number by 6.

Example: Find $16\frac{2}{3}\%$ of 90.

Solution: $90 \div 6 = 15$; or $90 \times \frac{1}{6} = 15$, answer

Proving the answer: What per cent is 15 of 90?

$$\frac{15}{90} = \frac{1}{6}; \times 100 = 16\frac{2}{3}\%$$

Other examples: $16\frac{2}{3}\%$ of $54 = 9$

$16\frac{2}{3}\%$ of $7.50 = \$1.25$

6-24. To Find 20% of a Number

Divide the number by 5.

Example: Find 20% of 95.

Solution: $95 \div 5 = 19$; or $95 \times \frac{1}{5} = 19$, answer

Proving the answer: What per cent is 19 of 95?

$$\frac{19}{95} = \frac{1}{5}; \times 100 = 20\%$$

Other examples: 20% of $65 = 13$

20% of $4.50 = \$.90$

6-25. To Find $33\frac{1}{3}\%$ of a Number

Divide the number by 3.

Example: Find $33\frac{1}{3}\%$ of 87.

Solution: $87 \div 3 = 29$; or $87 \times \frac{1}{3} = 29$, answer

Proving the answer: What per cent is 29 of 87?

$$\frac{29}{87} = \frac{1}{3}; \times 100 = 33\frac{1}{3}\%$$

Other examples: $33\frac{1}{3}\%$ of $72 = 24$

$33\frac{1}{3}\%$ of $10.50 = \$3.50$

6-26. To Find 50% of a Number

Divide the number by 2.

Example: Find 50% of 96.

Solution: $96 \div 2 = 48$; or $96 \times \frac{1}{2} = 48$, answer

Proving the answer: What per cent is 48 of 96%?

$$\frac{48}{96} = \frac{1}{2}; \times 100 = 50\%$$

Other examples: 50% of $185 = 92\frac{1}{2}$

50% of $3.50 = \$1.75$

6-27. To Find $66\frac{2}{3}\%$ of a Number

Divide the number by 3 and multiply by 2.

Example: Find $66\frac{2}{3}\%$ of 108

Solution: $108 \div 3 = 36; \times 2 = 72$, answer

or: $108 \times \frac{2}{3} = 72$

Proving the answer: What per cent is 72 of 108?

$$\frac{72}{108} = \frac{2}{3}; \times 100 = 66\frac{2}{3}\%$$

Other examples: $66\frac{2}{3}\%$ of $45 = 30$

$66\frac{2}{3}\%$ of $\$4.86 = \$3.24$

Table for Finding Certain Per Cents

To Find This Per Cent of a Number (N)	Method
$2\frac{1}{2}\%$	N/40
$12\frac{1}{2}\%$	N/8
$16\frac{2}{3}\%$	N/6
20%	N/5
25%	N/4
$33\frac{1}{3}\%$	N/3
$37\frac{1}{2}\%$	N/8 × 3
50%	N/2
$62\frac{1}{2}\%$	N/8 × 5
$66\frac{2}{3}\%$	N/3 × 2
$66\frac{2}{3}\%$	N − N/3
75%	N × 3/4
75%	N − N/4
$87\frac{1}{2}\%$	N − N/8

INDEX

A CATALOG OF SELECTED
DOVER BOOKS
IN ALL FIELDS OF INTEREST

A CATALOG OF SELECTED DOVER
BOOKS IN ALL FIELDS OF INTEREST

CONCERNING THE SPIRITUAL IN ART, Wassily Kandinsky. Pioneering work by father of abstract art. Thoughts on color theory, nature of art. Analysis of earlier masters. 12 illustrations. 80pp. of text. 5⅜ x 8½. 0-486-23411-8

CELTIC ART: The Methods of Construction, George Bain. Simple geometric techniques for making Celtic interlacements, spirals, Kells-type initials, animals, humans, etc. Over 500 illustrations. 160pp. 9 x 12. (Available in U.S. only.) 0-486-22923-8

AN ATLAS OF ANATOMY FOR ARTISTS, Fritz Schider. Most thorough reference work on art anatomy in the world. Hundreds of illustrations, including selections from works by Vesalius, Leonardo, Goya, Ingres, Michelangelo, others. 593 illustrations. 192pp. 7⅛ x 10¼. 0-486-20241-0

CELTIC HAND STROKE-BY-STROKE (Irish Half-Uncial from "The Book of Kells"): An Arthur Baker Calligraphy Manual, Arthur Baker. Complete guide to creating each letter of the alphabet in distinctive Celtic manner. Covers hand position, strokes, pens, inks, paper, more. Illustrated. 48pp. 8¼ x 11. 0-486-24336-2

EASY ORIGAMI, John Montroll. Charming collection of 32 projects (hat, cup, pelican, piano, swan, many more) specially designed for the novice origami hobbyist. Clearly illustrated easy-to-follow instructions insure that even beginning papercrafters will achieve successful results. 48pp. 8¼ x 11. 0-486-27298-2

BLOOMINGDALE'S ILLUSTRATED 1886 CATALOG: Fashions, Dry Goods and Housewares, Bloomingdale Brothers. Famed merchants' extremely rare catalog depicting about 1,700 products: clothing, housewares, firearms, dry goods, jewelry, more. Invaluable for dating, identifying vintage items. Also, copyright-free graphics for artists, designers. Co-published with Henry Ford Museum & Greenfield Village. 160pp. 8¼ x 11. 0-486-25780-0

THE ART OF WORLDLY WISDOM, Baltasar Gracian. "Think with the few and speak with the many," "Friends are a second existence," and "Be able to forget" are among this 1637 volume's 300 pithy maxims. A perfect source of mental and spiritual refreshment, it can be opened at random and appreciated either in brief or at length. 128pp. 5⅜ x 8½. 0-486-44034-6

JOHNSON'S DICTIONARY: A Modern Selection, Samuel Johnson (E. L. McAdam and George Milne, eds.). This modern version reduces the original 1755 edition's 2,300 pages of definitions and literary examples to a more manageable length, retaining the verbal pleasure and historical curiosity of the original. 480pp. 5³⁄₁₆ x 8¼. 0-486-44089-3

ADVENTURES OF HUCKLEBERRY FINN, Mark Twain, Illustrated by E. W. Kemble. A work of eternal richness and complexity, a source of ongoing critical debate, and a literary landmark, Twain's 1885 masterpiece about a barefoot boy's journey of self-discovery has enthralled readers around the world. This handsome clothbound reproduction of the first edition features all 174 of the original black-and-white illustrations. 368pp. 5⅜ x 8½. 0-486-44322-1

THE CLARINET AND CLARINET PLAYING, David Pino. Lively, comprehensive work features suggestions about technique, musicianship, and musical interpretation, as well as guidelines for teaching, making your own reeds, and preparing for public performance. Includes an intriguing look at clarinet history. "A godsend," *The Clarinet,* Journal of the International Clarinet Society. Appendixes. 7 illus. 320pp. 5⅜ x 8½. 0-486-40270-3

HOLLYWOOD GLAMOR PORTRAITS, John Kobal (ed.). 145 photos from 1926-49. Harlow, Gable, Bogart, Bacall; 94 stars in all. Full background on photographers, technical aspects. 160pp. 8⅜ x 11¼. 0-486-23352-9

THE RAVEN AND OTHER FAVORITE POEMS, Edgar Allan Poe. Over 40 of the author's most memorable poems: "The Bells," "Ulalume," "Israfel," "To Helen," "The Conqueror Worm," "Eldorado," "Annabel Lee," many more. Alphabetic lists of titles and first lines. 64pp. 5³⁄₁₆ x 8¼. 0-486-26685-0

PERSONAL MEMOIRS OF U. S. GRANT, Ulysses Simpson Grant. Intelligent, deeply moving firsthand account of Civil War campaigns, considered by many the finest military memoirs ever written. Includes letters, historic photographs, maps and more. 528pp. 6⅛ x 9¼. 0-486-28587-1

POE ILLUSTRATED: Art by Doré, Dulac, Rackham and Others, selected and edited by Jeff A. Menges. More than 100 compelling illustrations, in brilliant color and crisp black-and-white, include scenes from "The Raven," "The Pit and the Pendulum," "The Gold-Bug," and other stories and poems. 96pp. 8⅜ x 11. 0-486-45746-X

RUSSIAN STORIES/RUSSKIE RASSKAZY: A Dual-Language Book, edited by Gleb Struve. Twelve tales by such masters as Chekhov, Tolstoy, Dostoevsky, Pushkin, others. Excellent word-for-word English translations on facing pages, plus teaching and study aids, Russian/English vocabulary, biographical/critical introductions, more. 416pp. 5⅜ x 8½. 0-486-26244-8

PHILADELPHIA THEN AND NOW: 60 Sites Photographed in the Past and Present, Kenneth Finkel and Susan Oyama. Rare photographs of City Hall, Logan Square, Independence Hall, Betsy Ross House, other landmarks juxtaposed with contemporary views. Captures changing face of historic city. Introduction. Captions. 128pp. 8¼ x 11. 0-486-25790-8

NORTH AMERICAN INDIAN LIFE: Customs and Traditions of 23 Tribes, Elsie Clews Parsons (ed.). 27 fictionalized essays by noted anthropologists examine religion, customs, government, additional facets of life among the Winnebago, Crow, Zuni, Eskimo, other tribes. 480pp. 6⅛ x 9¼. 0-486-27377-6

TECHNICAL MANUAL AND DICTIONARY OF CLASSICAL BALLET, Gail Grant. Defines, explains, comments on steps, movements, poses and concepts. 15-page pictorial section. Basic book for student, viewer. 127pp. 5⅜ x 8½. 0-486-21843-0

THE MALE AND FEMALE FIGURE IN MOTION: 60 Classic Photographic Sequences, Eadweard Muybridge. 60 true-action photographs of men and women walking, running, climbing, bending, turning, etc., reproduced from a rare 19th-century masterpiece. vi + 121pp. 9 x 12. 0-486-24745-7

CATALOG OF DOVER BOOKS

ANIMALS: 1,419 Copyright-Free Illustrations of Mammals, Birds, Fish, Insects, etc., Jim Harter (ed.). Clear wood engravings present, in extremely lifelike poses, over 1,000 species of animals. One of the most extensive pictorial sourcebooks of its kind. Captions. Index. 284pp. 9 x 12. 0-486-23766-4

1001 QUESTIONS ANSWERED ABOUT THE SEASHORE, N. J. Berrill and Jacquelyn Berrill. Queries answered about dolphins, sea snails, sponges, starfish, fishes, shore birds, many others. Covers appearance, breeding, growth, feeding, much more. 305pp. 5¼ x 8¼. 0-486-23366-9

ATTRACTING BIRDS TO YOUR YARD, William J. Weber. Easy-to-follow guide offers advice on how to attract the greatest diversity of birds: birdhouses, feeders, water and waterers, much more. 96pp. 5³⁄₁₆ x 8¼. 0-486-28927-3

MEDICINAL AND OTHER USES OF NORTH AMERICAN PLANTS: A Historical Survey with Special Reference to the Eastern Indian Tribes, Charlotte Erichsen-Brown. Chronological historical citations document 500 years of usage of plants, trees, shrubs native to eastern Canada, northeastern U.S. Also complete identifying information. 343 illustrations. 544pp. 6½ x 9¼. 0-486-25951-X

STORYBOOK MAZES, Dave Phillips. 23 stories and mazes on two-page spreads: Wizard of Oz, Treasure Island, Robin Hood, etc. Solutions. 64pp. 8¼ x 11. 0-486-23628-5

AMERICAN NEGRO SONGS: 230 Folk Songs and Spirituals, Religious and Secular, John W. Work. This authoritative study traces the African influences of songs sung and played by black Americans at work, in church, and as entertainment. The author discusses the lyric significance of such songs as "Swing Low, Sweet Chariot," "John Henry," and others and offers the words and music for 230 songs. Bibliography. Index of Song Titles. 272pp. 6½ x 9¼. 0-486-40271-1

MOVIE-STAR PORTRAITS OF THE FORTIES, John Kobal (ed.). 163 glamor, studio photos of 106 stars of the 1940s: Rita Hayworth, Ava Gardner, Marlon Brando, Clark Gable, many more. 176pp. 8⅜ x 11¼. 0-486-23546-7

YEKL and THE IMPORTED BRIDEGROOM AND OTHER STORIES OF YIDDISH NEW YORK, Abraham Cahan. Film Hester Street based on *Yekl* (1896). Novel, other stories among first about Jewish immigrants on N.Y.'s East Side. 240pp. 5⅜ x 8½. 0-486-22427-9

SELECTED POEMS, Walt Whitman. Generous sampling from *Leaves of Grass*. Twenty-four poems include "I Hear America Singing," "Song of the Open Road," "I Sing the Body Electric," "When Lilacs Last in the Dooryard Bloom'd," "O Captain! My Captain!"—all reprinted from an authoritative edition. Lists of titles and first lines. 128pp. 5³⁄₁₆ x 8¼. 0-486-26878-0

SONGS OF EXPERIENCE: Facsimile Reproduction with 26 Plates in Full Color, William Blake. 26 full-color plates from a rare 1826 edition. Includes "The Tyger," "London," "Holy Thursday," and other poems. Printed text of poems. 48pp. 5¼ x 7. 0-486-24636-1

THE BEST TALES OF HOFFMANN, E. T. A. Hoffmann. 10 of Hoffmann's most important stories: "Nutcracker and the King of Mice," "The Golden Flowerpot," etc. 458pp. 5⅜ x 8½. 0-486-21793-0

THE BOOK OF TEA, Kakuzo Okakura. Minor classic of the Orient: entertaining, charming explanation, interpretation of traditional Japanese culture in terms of tea ceremony. 94pp. 5⅜ x 8½. 0-486-20070-1

CATALOG OF DOVER BOOKS

FRENCH STORIES/CONTES FRANÇAIS: A Dual-Language Book, Wallace Fowlie. Ten stories by French masters, Voltaire to Camus: "Micromegas" by Voltaire; "The Atheist's Mass" by Balzac; "Minuet" by de Maupassant; "The Guest" by Camus, six more. Excellent English translations on facing pages. Also French-English vocabulary list, exercises, more. 352pp. 5⅜ x 8½. 0-486-26443-2

CHICAGO AT THE TURN OF THE CENTURY IN PHOTOGRAPHS: 122 Historic Views from the Collections of the Chicago Historical Society, Larry A. Viskochil. Rare large-format prints offer detailed views of City Hall, State Street, the Loop, Hull House, Union Station, many other landmarks, circa 1904-1913. Introduction. Captions. Maps. 144pp. 9⅜ x 12¼. 0-486-24656-6

OLD BROOKLYN IN EARLY PHOTOGRAPHS, 1865-1929, William Lee Younger. Luna Park, Gravesend race track, construction of Grand Army Plaza, moving of Hotel Brighton, etc. 157 previously unpublished photographs. 165pp. 8⅞ x 11¾. 0-486-23587-4

THE MYTHS OF THE NORTH AMERICAN INDIANS, Lewis Spence. Rich anthology of the myths and legends of the Algonquins, Iroquois, Pawnees and Sioux, prefaced by an extensive historical and ethnological commentary. 36 illustrations. 480pp. 5⅜ x 8½. 0-486-25967-6

AN ENCYCLOPEDIA OF BATTLES: Accounts of Over 1,560 Battles from 1479 B.C. to the Present, David Eggenberger. Essential details of every major battle in recorded history from the first battle of Megiddo in 1479 B.C. to Grenada in 1984. List of Battle Maps. New Appendix covering the years 1967-1984. Index. 99 illustrations. 544pp. 6½ x 9¼. 0-486-24913-1

SAILING ALONE AROUND THE WORLD, Captain Joshua Slocum. First man to sail around the world, alone, in small boat. One of the great feats of seamanship told in delightful manner. 67 illustrations. 294pp. 5⅜ x 8½. 0-486-20326-3

ANARCHISM AND OTHER ESSAYS, Emma Goldman. Powerful, penetrating, prophetic essays on direct action, role of minorities, prison reform, puritan hypocrisy, violence, etc. 271pp. 5⅜ x 8½. 0-486-22484-8

MYTHS OF THE HINDUS AND BUDDHISTS, Ananda K. Coomaraswamy and Sister Nivedita. Great stories of the epics; deeds of Krishna, Shiva, taken from puranas, Vedas, folk tales; etc. 32 illustrations. 400pp. 5⅜ x 8½. 0-486-21759-0

MY BONDAGE AND MY FREEDOM, Frederick Douglass. Born a slave, Douglass became outspoken force in antislavery movement. The best of Douglass' autobiographies. Graphic description of slave life. 464pp. 5⅜ x 8½. 0-486-22457-0

FOLLOWING THE EQUATOR: A Journey Around the World, Mark Twain. Fascinating humorous account of 1897 voyage to Hawaii, Australia, India, New Zealand, etc. Ironic, bemused reports on peoples, customs, climate, flora and fauna, politics, much more. 197 illustrations. 720pp. 5⅜ x 8½. 0-486-26113-1

GREAT SPEECHES BY AMERICAN WOMEN, edited by James Daley. Here are 21 legendary speeches from the country's most inspirational female voices, including Sojourner Truth, Susan B. Anthony, Eleanor Roosevelt, Hillary Rodham Clinton, Nancy Pelosi, and many others. 192pp. 5³⁄₁₆ x 8¼. 0-486-46141-6

THE MYTHS OF GREECE AND ROME, H. A. Guerber. A classic of mythology, generously illustrated, long prized for its simple, graphic, accurate retelling of the principal myths of Greece and Rome, and for its commentary on their origins and significance. With 64 illustrations by Michelangelo, Raphael, Titian, Rubens, Canova, Bernini and others. 480pp. 5⅜ x 8½. 0-486-27584-1

CATALOG OF DOVER BOOKS

HOW TO DO BEADWORK, Mary White. Fundamental book on craft from simple projects to five-bead chains and woven works. 106 illustrations. 142pp. 5⅜ x 8.
0-486-20697-1

THE 1912 AND 1915 GUSTAV STICKLEY FURNITURE CATALOGS, Gustav Stickley. With over 200 detailed illustrations and descriptions, these two catalogs are essential reading and reference materials and identification guides for Stickley furniture. Captions cite materials, dimensions and prices. 112pp. 6½ x 9¼. 0-486-26676-1

SIX GREAT DIALOGUES: Apology, Crito, Phaedo, Phaedrus, Symposium, The Republic, Plato, translated by Benjamin Jowett. Plato's Dialogues rank among Western civilization's most important and influential philosophical works. These 6 selections of his major works explore a broad range of enduringly relevant issues. Authoritative Jowett translations. 480pp. 5³⁄₁₆ x 8¼. 0-486-45465-7

DEMONOLATRY: An Account of the Historical Practice of Witchcraft, Nicolas Remy, edited with an Introduction and Notes by Montague Summers, translated by E. A. Ashwin. This extremely influential 1595 study was frequently cited at witchcraft trials. In addition to lurid details of satanic pacts and sexual perversity, it presents the particulars of numerous court cases. 240pp. 6½ x 9¼. 0-486-46137-8

VICTORIAN FASHIONS AND COSTUMES FROM HARPER'S BAZAAR, 1867–1898, Stella Blum (ed.). Day costumes, evening wear, sports clothes, shoes, hats, other accessories in over 1,000 detailed engravings. 320pp. 9⅜ x 12¼.
0-486-22990-4

THE LONG ISLAND RAIL ROAD IN EARLY PHOTOGRAPHS, Ron Ziel. Over 220 rare photos, informative text document origin (1844) and development of rail service on Long Island. Vintage views of early trains, locomotives, stations, passengers, crews, much more. Captions. 8⅞ x 11¾. 0-486-26301-0

VOYAGE OF THE LIBERDADE, Joshua Slocum. Great 19th-century mariner's thrilling, first-hand account of the wreck of his ship off South America, the 35-foot boat he built from the wreckage, and its remarkable voyage home. 128pp. 5⅜ x 8½.
0-486-40022-0

TEN BOOKS ON ARCHITECTURE, Vitruvius. The most important book ever written on architecture. Early Roman aesthetics, technology, classical orders, site selection, all other aspects. Morgan translation. 331pp. 5⅜ x 8½. 0-486-20645-9

THE HUMAN FIGURE IN MOTION, Eadweard Muybridge. More than 4,500 stopped-action photos, in action series, showing undraped men, women, children jumping, lying down, throwing, sitting, wrestling, carrying, etc. 390pp. 7⅞ x 10⅝.
0-486-20204-6 Clothbd.

TREES OF THE EASTERN AND CENTRAL UNITED STATES AND CANADA, William M. Harlow. Best one-volume guide to 140 trees. Full descriptions, woodlore, range, etc. Over 600 illustrations. Handy size. 288pp. 4½ x 6⅜. 0-486-20395-6

MY FIRST BOOK OF TCHAIKOVSKY: Favorite Pieces in Easy Piano Arrangements, edited by David Dutkanicz. These special arrangements of favorite Tchaikovsky themes are ideal for beginner pianists, child or adult. Contents include themes from "The Nutcracker," "March Slav," Symphonies Nos. 5 and 6, "Swan Lake," "Sleeping Beauty," and more. 48pp. 8¼ x 11. 0-486-46416-4

BIG BOOK OF MAZES AND LABYRINTHS, Walter Shepherd. 50 mazes and labyrinths in all–classical, solid, ripple, and more–in one great volume. Perfect inexpensive puzzler for clever youngsters. Full solutions. 112pp. 8⅛ x 11. 0-486-22951-3

PIANO TUNING, J. Cree Fischer. Clearest, best book for beginner, amateur. Simple repairs, raising dropped notes, tuning by easy method of flattened fifths. No previous skills needed. 4 illustrations. 201pp. 5⅜ x 8½. 0-486-23267-0

CATALOG OF DOVER BOOKS

HINTS TO SINGERS, Lillian Nordica. Selecting the right teacher, developing confidence, overcoming stage fright, and many other important skills receive thoughtful discussion in this indispensible guide, written by a world-famous diva of four decades' experience. 96pp. 5⅜ x 8½. 0-486-40094-8

THE COMPLETE NONSENSE OF EDWARD LEAR, Edward Lear. All nonsense limericks, zany alphabets, Owl and Pussycat, songs, nonsense botany, etc., illustrated by Lear. Total of 320pp. 5⅜ x 8½. (Available in U.S. only.) 0-486-20167-8

VICTORIAN PARLOUR POETRY: An Annotated Anthology, Michael R. Turner. 117 gems by Longfellow, Tennyson, Browning, many lesser-known poets. "The Village Blacksmith," "Curfew Must Not Ring Tonight," "Only a Baby Small," dozens more, often difficult to find elsewhere. Index of poets, titles, first lines. xxiii + 325pp. 5⅜ x 8¼. 0-486-27044-0

DUBLINERS, James Joyce. Fifteen stories offer vivid, tightly focused observations of the lives of Dublin's poorer classes. At least one, "The Dead," is considered a masterpiece. Reprinted complete and unabridged from standard edition. 160pp. 5³⁄₁₆ x 8¼. 0-486-26870-5

THE LITTLE RED SCHOOLHOUSE, Eric Sloane. Harkening back to a time when the three Rs stood for reading, 'riting, and religion, Sloane's sketchbook explores the history of early American schools. Includes marvelous illustrations of one-room New England schoolhouses, desks, and benches. 48pp. 8¼ x 11. 0-486-45604-8

THE BOOK OF THE SACRED MAGIC OF ABRAMELIN THE MAGE, translated by S. MacGregor Mathers. Medieval manuscript of ceremonial magic. Basic document in Aleister Crowley, Golden Dawn groups. 268pp. 5⅜ x 8½. 0-486-23211-5

THE BATTLES THAT CHANGED HISTORY, Fletcher Pratt. Eminent historian profiles 16 crucial conflicts, ancient to modern, that changed the course of civilization. 352pp. 5⅜ x 8½. 0-486-41129-X

NEW RUSSIAN-ENGLISH AND ENGLISH-RUSSIAN DICTIONARY, M. A. O'Brien. This is a remarkably handy Russian dictionary, containing a surprising amount of information, including over 70,000 entries. 366pp. 4½ x 6⅛. 0-486-20208-9

NEW YORK IN THE FORTIES, Andreas Feininger. 162 brilliant photographs by the well-known photographer, formerly with *Life* magazine. Commuters, shoppers, Times Square at night, much else from city at its peak. Captions by John von Hartz. 181pp. 9¼ x 10¾. 0-486-23585-8

INDIAN SIGN LANGUAGE, William Tomkins. Over 525 signs developed by Sioux and other tribes. Written instructions and diagrams. Also 290 pictographs. 111pp. 6⅛ x 9¼. 0-486-22029-X

ANATOMY: A Complete Guide for Artists, Joseph Sheppard. A master of figure drawing shows artists how to render human anatomy convincingly. Over 460 illustrations. 224pp. 8⅜ x 11¼. 0-486-27279-6

MEDIEVAL CALLIGRAPHY: Its History and Technique, Marc Drogin. Spirited history, comprehensive instruction manual covers 13 styles (ca. 4th century through 15th). Excellent photographs; directions for duplicating medieval techniques with modern tools. 224pp. 8⅜ x 11¼. 0-486-26142-5

CATALOG OF DOVER BOOKS

DRIED FLOWERS: How to Prepare Them, Sarah Whitlock and Martha Rankin. Complete instructions on how to use silica gel, meal and borax, perlite aggregate, sand and borax, glycerine and water to create attractive permanent flower arrangements. 12 illustrations. 32pp. 5⅜ x 8½. 0-486-21802-3

EASY-TO-MAKE BIRD FEEDERS FOR WOODWORKERS, Scott D. Campbell. Detailed, simple-to-use guide for designing, constructing, caring for and using feeders. Text, illustrations for 12 classic and contemporary designs. 96pp. 5⅜ x 8½. 0-486-25847-5

THE COMPLETE BOOK OF BIRDHOUSE CONSTRUCTION FOR WOODWORKERS, Scott D. Campbell. Detailed instructions, illustrations, tables. Also data on bird habitat and instinct patterns. Bibliography. 3 tables. 63 illustrations in 15 figures. 48pp. 5¼ x 8½. 0-486-24407-5

SCOTTISH WONDER TALES FROM MYTH AND LEGEND, Donald A. Mackenzie. 16 lively tales tell of giants rumbling down mountainsides, of a magic wand that turns stone pillars into warriors, of gods and goddesses, evil hags, powerful forces and more. 240pp. 5⅜ x 8½. 0-486-29677-6

THE HISTORY OF UNDERCLOTHES, C. Willett Cunnington and Phyllis Cunnington. Fascinating, well-documented survey covering six centuries of English undergarments, enhanced with over 100 illustrations: 12th-century laced-up bodice, footed long drawers (1795), 19th-century bustles, l9th-century corsets for men, Victorian "bust improvers," much more. 272pp. 5⅜ x 8¼. 0-486-27124-2

FIRST FRENCH READER: A Beginner's Dual-Language Book, edited and translated by Stanley Appelbaum. This anthology introduces fifty legendary writers—Voltaire, Balzac, Baudelaire, Proust, more—through passages from The Red and the Black, Les Misérables, Madame Bovary, and other classics. Original French text plus English translation on facing pages. 240pp. 5⅜ x 8½. 0-486-46178-5

WILBUR AND ORVILLE: A Biography of the Wright Brothers, Fred Howard. Definitive, crisply written study tells the full story of the brothers' lives and work. A vividly written biography, unparalleled in scope and color, that also captures the spirit of an extraordinary era. 560pp. 6⅛ x 9¼. 0-486-40297-5

THE ARTS OF THE SAILOR: Knotting, Splicing and Ropework, Hervey Garrett Smith. Indispensable shipboard reference covers tools, basic knots and useful hitches; handsewing and canvas work, more. Over 100 illustrations. Delightful reading for sea lovers. 256pp. 5⅜ x 8½. 0-486-26440-8

FRANK LLOYD WRIGHT'S FALLINGWATER: The House and Its History, Second, Revised Edition, Donald Hoffmann. A total revision—both in text and illustrations—of the standard document on Fallingwater, the boldest, most personal architectural statement of Wright's mature years, updated with valuable new material from the recently opened Frank Lloyd Wright Archives. "Fascinating"—The New York Times. 116 illustrations. 128pp. 9¼ x 10¾. 0-486-27430-6

PHOTOGRAPHIC SKETCHBOOK OF THE CIVIL WAR, Alexander Gardner. 100 photos taken on field during the Civil War. Famous shots of Manassas Harper's Ferry, Lincoln, Richmond, slave pens, etc. 244pp. 10⅝ x 8¼. 0-486-22731-6

FIVE ACRES AND INDEPENDENCE, Maurice G. Kains. Great back-to-the-land classic explains basics of self-sufficient farming. The one book to get. 95 illustrations. 397pp. 5⅜ x 8½. 0-486-20974-1

THE MALLEUS MALEFICARUM OF KRAMER AND SPRENGER, translated by Montague Summers. Full text of most important witchhunter's "bible," used by both Catholics and Protestants. 278pp. 6⅝ x 10. 0-486-22802-9

SPANISH STORIES/CUENTOS ESPAÑOLES: A Dual-Language Book, Angel Flores (ed.). Unique format offers 13 great stories in Spanish by Cervantes, Borges, others. Faithful English translations on facing pages. 352pp. 5⅜ x 8½.
0-486-25399-6

GARDEN CITY, LONG ISLAND, IN EARLY PHOTOGRAPHS, 1869–1919, Mildred H. Smith. Handsome treasury of 118 vintage pictures, accompanied by carefully researched captions, document the Garden City Hotel fire (1899), the Vanderbilt Cup Race (1908), the first airmail flight departing from the Nassau Boulevard Aerodrome (1911), and much more. 96pp. 8⅞ x 11¾. 0-486-40669-5

OLD QUEENS, N.Y., IN EARLY PHOTOGRAPHS, Vincent F. Seyfried and William Asadorian. Over 160 rare photographs of Maspeth, Jamaica, Jackson Heights, and other areas. Vintage views of DeWitt Clinton mansion, 1939 World's Fair and more. Captions. 192pp. 8⅞ x 11. 0-486-26358-4

CAPTURED BY THE INDIANS: 15 Firsthand Accounts, 1750-1870, Frederick Drimmer. Astounding true historical accounts of grisly torture, bloody conflicts, relentless pursuits, miraculous escapes and more, by people who lived to tell the tale. 384pp. 5⅜ x 8½. 0-486-24901-8

THE WORLD'S GREAT SPEECHES (Fourth Enlarged Edition), Lewis Copeland, Lawrence W. Lamm, and Stephen J. McKenna. Nearly 300 speeches provide public speakers with a wealth of updated quotes and inspiration–from Pericles' funeral oration and William Jennings Bryan's "Cross of Gold Speech" to Malcolm X's powerful words on the Black Revolution and Earl of Spenser's tribute to his sister, Diana, Princess of Wales. 944pp. 5⅜ x 8⅜. 0-486-40903-1

THE BOOK OF THE SWORD, Sir Richard F. Burton. Great Victorian scholar/adventurer's eloquent, erudite history of the "queen of weapons"–from prehistory to early Roman Empire. Evolution and development of early swords, variations (sabre, broadsword, cutlass, scimitar, etc.), much more. 336pp. 6⅛ x 9¼.
0-486-25434-8

AUTOBIOGRAPHY: The Story of My Experiments with Truth, Mohandas K. Gandhi. Boyhood, legal studies, purification, the growth of the Satyagraha (nonviolent protest) movement. Critical, inspiring work of the man responsible for the freedom of India. 480pp. 5⅜ x 8½. (Available in U.S. only.) 0-486-24593-4

CELTIC MYTHS AND LEGENDS, T. W. Rolleston. Masterful retelling of Irish and Welsh stories and tales. Cuchulain, King Arthur, Deirdre, the Grail, many more. First paperback edition. 58 full-page illustrations. 512pp. 5⅜ x 8½. 0-486-26507-2

THE PRINCIPLES OF PSYCHOLOGY, William James. Famous long course complete, unabridged. Stream of thought, time perception, memory, experimental methods; great work decades ahead of its time. 94 figures. 1,391pp. 5⅜ x 8½. 2-vol. set.
Vol. I: 0-486-20381-6   Vol. II: 0-486-20382-4

THE WORLD AS WILL AND REPRESENTATION, Arthur Schopenhauer. Definitive English translation of Schopenhauer's life work, correcting more than 1,000 errors, omissions in earlier translations. Translated by E. F. J. Payne. Total of 1,269pp. 5⅜ x 8½. 2-vol. set.   Vol. 1: 0-486-21761-2   Vol. 2: 0-486-21762-0

CATALOG OF DOVER BOOKS

MAKING FURNITURE MASTERPIECES: 30 Projects with Measured Drawings, Franklin H. Gottshall. Step-by-step instructions, illustrations for constructing handsome, useful pieces, among them a Sheraton desk, Chippendale chair, Spanish desk, Queen Anne table and a William and Mary dressing mirror. 224pp. 8⅛ x 11¼.
0-486-29338-6

NORTH AMERICAN INDIAN DESIGNS FOR ARTISTS AND CRAFTSPEOPLE, Eva Wilson. Over 360 authentic copyright-free designs adapted from Navajo blankets, Hopi pottery, Sioux buffalo hides, more. Geometrics, symbolic figures, plant and animal motifs, etc. 128pp. 8⅜ x 11. (Not for sale in the United Kingdom.) 0-486-25341-4

THE FOSSIL BOOK: A Record of Prehistoric Life, Patricia V. Rich et al. Profusely illustrated definitive guide covers everything from single-celled organisms and dinosaurs to birds and mammals and the interplay between climate and man. Over 1,500 illustrations. 760pp. 7½ x 10⅛. 0-486-29371-8

VICTORIAN ARCHITECTURAL DETAILS: Designs for Over 700 Stairs, Mantels, Doors, Windows, Cornices, Porches, and Other Decorative Elements, A. J. Bicknell & Company. Everything from dormer windows and piazzas to balconies and gable ornaments. Also includes elevations and floor plans for handsome, private residences and commercial structures. 80pp. 9⅜ x 12¼. 0-486-44015-X

WESTERN ISLAMIC ARCHITECTURE: A Concise Introduction, John D. Hoag. Profusely illustrated critical appraisal compares and contrasts Islamic mosques and palaces—from Spain and Egypt to other areas in the Middle East. 139 illustrations. 128pp. 6 x 9. 0-486-43760-4

CHINESE ARCHITECTURE: A Pictorial History, Liang Ssu-ch'eng. More than 240 rare photographs and drawings depict temples, pagodas, tombs, bridges, and imperial palaces comprising much of China's architectural heritage. 152 halftones, 94 diagrams. 232pp. 10¾ x 9⅞. 0-486-43999-2

THE RENAISSANCE: Studies in Art and Poetry, Walter Pater. One of the most talked-about books of the 19th century, *The Renaissance* combines scholarship and philosophy in an innovative work of cultural criticism that examines the achievements of Botticelli, Leonardo, Michelangelo, and other artists. "The holy writ of beauty."—Oscar Wilde. 160pp. 5⅜ x 8½. 0-486-44025-7

A TREATISE ON PAINTING, Leonardo da Vinci. The great Renaissance artist's practical advice on drawing and painting techniques covers anatomy, perspective, composition, light and shadow, and color. A classic of art instruction, it features 48 drawings by Nicholas Poussin and Leon Battista Alberti. 192pp. 5⅜ x 8½.
0-486-44155-5

THE ESSENTIAL JEFFERSON, Thomas Jefferson, edited by John Dewey. This extraordinary primer offers a superb survey of Jeffersonian thought. It features writings on political and economic philosophy, morals and religion, intellectual freedom and progress, education, secession, slavery, and more. 176pp. 5⅜ x 8½.
0-486-46599-3

WASHINGTON IRVING'S RIP VAN WINKLE, Illustrated by Arthur Rackham. Lovely prints that established artist as a leading illustrator of the time and forever etched into the popular imagination a classic of Catskill lore. 51 full-color plates. 80pp. 8⅜ x 11. 0-486-44242-X

HENSCHE ON PAINTING, John W. Robichaux. Basic painting philosophy and methodology of a great teacher, as expounded in his famous classes and workshops on Cape Cod. 7 illustrations in color on covers. 80pp. 5⅜ x 8½. 0-486-43728-0

CATALOG OF DOVER BOOKS

LIGHT AND SHADE: A Classic Approach to Three-Dimensional Drawing, Mrs. Mary P. Merrifield. Handy reference clearly demonstrates principles of light and shade by revealing effects of common daylight, sunshine, and candle or artificial light on geometrical solids. 13 plates. 64pp. $5\frac{3}{8} \times 8\frac{1}{2}$. 0-486-44143-1

ASTROLOGY AND ASTRONOMY: A Pictorial Archive of Signs and Symbols, Ernst and Johanna Lehner. Treasure trove of stories, lore, and myth, accompanied by more than 300 rare illustrations of planets, the Milky Way, signs of the zodiac, comets, meteors, and other astronomical phenomena. 192pp. $8\frac{3}{8} \times 11$. 0-486-43981-X

JEWELRY MAKING: Techniques for Metal, Tim McCreight. Easy-to-follow instructions and carefully executed illustrations describe tools and techniques, use of gems and enamels, wire inlay, casting, and other topics. 72 line illustrations and diagrams. 176pp. $8\frac{1}{4} \times 10\frac{7}{8}$. 0-486-44043-5

MAKING BIRDHOUSES: Easy and Advanced Projects, Gladstone Califf. Easy-to-follow instructions include diagrams for everything from a one-room house for bluebirds to a forty-two-room structure for purple martins. 56 plates; 4 figures. 80pp. $8\frac{3}{4} \times 6\frac{3}{8}$. 0-486-44183-0

LITTLE BOOK OF LOG CABINS: How to Build and Furnish Them, William S. Wicks. Handy how-to manual, with instructions and illustrations for building cabins in the Adirondack style, fireplaces, stairways, furniture, beamed ceilings, and more. 102 line drawings. 96pp. $8\frac{3}{4} \times 6\frac{3}{8}$. 0-486-44259-4

THE SEASONS OF AMERICA PAST, Eric Sloane. From "sugaring time" and strawberry picking to Indian summer and fall harvest, a whole year's activities described in charming prose and enhanced with 79 of the author's own illustrations. 160pp. $8\frac{1}{4} \times 11$. 0-486-44220-9

THE METROPOLIS OF TOMORROW, Hugh Ferriss. Generous, prophetic vision of the metropolis of the future, as perceived in 1929. Powerful illustrations of towering structures, wide avenues, and rooftop parks–all features in many of today's modern cities. 59 illustrations. 144pp. $8\frac{1}{4} \times 11$. 0-486-43727-2

THE PATH TO ROME, Hilaire Belloc. This 1902 memoir abounds in lively vignettes from a vanished time, recounting a pilgrimage on foot across the Alps and Apennines in order to "see all Europe which the Christian Faith has saved." 77 of the author's original line drawings complement his sparkling prose. 272pp. $5\frac{3}{8} \times 8\frac{1}{2}$. 0-486-44001-X

THE HISTORY OF RASSELAS: Prince of Abissinia, Samuel Johnson. Distinguished English writer attacks eighteenth-century optimism and man's unrealistic estimates of what life has to offer. 112pp. $5\frac{3}{8} \times 8\frac{1}{2}$. 0-486-44094-X

A VOYAGE TO ARCTURUS, David Lindsay. A brilliant flight of pure fancy, where wild creatures crowd the fantastic landscape and demented torturers dominate victims with their bizarre mental powers. 272pp. $5\frac{3}{8} \times 8\frac{1}{2}$. 0-486-44198-9

Paperbound unless otherwise indicated. Available at your book dealer, online at **www.doverpublications.com**, or by writing to Dept. GI, Dover Publications, Inc., 31 East 2nd Street, Mineola, NY 11501. For current price information or for free catalogs (please indicate field of interest), write to Dover Publications or log on to **www.doverpublications.com** and see every Dover book in print. Dover publishes more than 400 books each year on science, elementary and advanced mathematics, biology, music, art, literary history, social sciences, and other areas.